T0241407

Synthesis Lectures on Visual Computing: Computer Graphics, Animation, Computational Photography and Imaging

Series Editor

Brian A. Barsky, University of California, Berkeley, Berkeley, USA

This series presents lectures on research and development in visual computing for an audience of professional developers, researchers, and advanced students. Topics of interest include computational photography, animation, visualization, special effects, game design, image techniques, computational geometry, modeling, rendering, and others of interest to the visual computing system developer or researcher.

Shiguang Liu

Image and Video Color Editing

 Springer

Shiguang Liu
College of Intelligence and Computing
Tianjin University
Tianjin, China

ISSN 2469-4215   ISSN 2469-4223   (electronic)
Synthesis Lectures on Visual Computing: Computer Graphics, Animation,
Computational Photography and Imaging
ISBN 978-3-031-26032-2   ISBN 978-3-031-26030-8   (eBook)
https://doi.org/10.1007/978-3-031-26030-8

This Springer imprint is published by the registered company Springer Nature Switzerland AG
The registered company address is: Gewerbestrasse 11, 6330 Cham, Switzerland

Preface

Digital image processing is generally processed by computer, so it is also called computer image processing. The purpose of early image processing is to improve the quality of an image so as to produce a better visual perception for the viewer. General image processing methods include image enhancement, restoration, coding, compression, etc. The first successful practical application was the Jet Propulsion Laboratory (JPL). They used image processing techniques such as geometric correction, gray scale transformation, noise removal and other methods to process thousands of lunar photos sent back by the space probe "Wanderer 7" in 1964, and successfully reconstructed the lunar surface by further considering the influence of the sun's position and the lunar environment. In the future aerospace technology, such as the exploration of Mars, Saturn and other planets, digital image processing technology has also played a critical role. Another great achievement of digital image processing is medicine. For example, CT (Computer Tomography) and non-invasive diagnostic technology have made epoch-making contributions to mankind. Video is a sequence of time varying images. Video processing is an extension of image processing, with an emphasis on both spatial and time coherence.

With the rapid development of digital image and video processing techniques, they have become the object of research among various disciplines in the fields of engineering, computer science, information science, statistics, physics, chemistry, biology, medicine and even social science. Nowadays, image and video processing technology has brought great economic and social benefits to mankind. In the near future, it will not only have more in-depth development in theory, but also be an indispensable and powerful tool in scientific research, social production and even human life in application.

One of the most common image and video processing methods is to change the color of an image (or video) to meet the user's demand. So far, there are many ways to change the color of an image (or video), including color transfer, colorization, decolorization, style transfer, enhancement, etc.

Image and video color transfer: Color transfer is one of the widely used color editing methods. It is a statistics-based technique. Its main idea is to transfer the color characteristics of a target image to another source image or video, so that the source image has the same color distribution as the target image. These methods focus on processing the color of an image or video to achieve some visual effect, ignoring the emotion an image or video conveys. To this end, some researchers developed emotional color transfer to transfer both the color and emotion of a target image to a source image (or video).

Image and video colorization: Colorization aims to adding colors to a grayscale image or video. This task is ill-posed in the sense that assigning the colors to a grayscale image without any prior knowledge is ambiguous. Most of the previous methods require some amount of user interventions to assist the coloring process. Moreover, video colorization technology is more difficult, since it also needs to keep the time-space consistency of the results. At present, there are a large number of grayscale or black and white images and video materials in various fields such as film and television, picture archives, medical treatment, etc. Coloring them can enhance the details and help people better identify and make better use of them.

Image and video decolorization: Decolorization refers to convert a color image or video into a grayscale one. This transformation process is a dimension reduction process, i.e., from a cubic matrix to a single matrix. In this process, information loss will inevitably occur. The key of decolorization is to make the decolorization results retain human's perception of the original color image or video as much as possible. It is widely used in daily life. For example, most newspapers choose to use grayscale images as illustrations to save expenses; on the other hand, grayscale images have become a favorite artistic choice for photographers around the world; in addition, grayscale images and videos are also widely used in medical imaging and surveillance.

Image and video style transfer: Given a source image or video and a style image, style transfer is to make the source image or video show the "style" of the style image. It can make an image or video display a variety of stylized effects, giving one more appealing visual appreciation. For example, it can convert a photo taken by yourself into an image with the artistic style of Monet, Van Gogh and other painting giants. Style transfer is closely related to texture synthesis. Texture synthesis attempts to capture the statistical relationship between pixels in the source image, while style transfer also preserves the content structure.

Image and video enhancement: Photography devices may capture images or videos with poor quality under a low light-level photographic environment. Due to the low visibility, these underexposed images and videos usually fail to present visually pleasing browsing. Image and video enhancement aim to reveal hidden details in underexposed videos and improve video quality, which can benefit the downstream image and video processing tasks. Image and video color enhancement technology has a wide range of applications, including target fusion, target tracking, video compression, etc. Enhancement of a low exposure video can also improve the appearance of surveillance video shot by a monitoring system under bad weather.

Most of these research results were published at leading venues in computer graphics, computer vision, image processing, multimedia, etc. This stimulated further research and helped establish a large research and development community. This monograph gives an overview of image and video color editing methods, with specific emphasis on fast techniques developed over the last 20 years.

Tianjin, China   Shiguang Liu
November 2022

Acknowledgments This work was supported in part by the National Natural Science Foundation of China under grant nos. 62072328 and 61672375. I will show thanks to my students Zhichao Song, Xiaoli Zhang, Yaxi Jiang, Min Pei, Huorong Luo, Ting Zhu, Xiang Zhang, Yu Zhang, Shichao Li and Hongchao Zhang for their excellent work during their study in Tianjin University.

Contents

About the Author

Shiguang Liu is currently a professor with the School of Computer Science and Technology, College of Intelligence and Computing, Tianjin University, P.R. China. He received a Ph.D. from the State Key Lab of CAD and CG, Zhejiang University, P.R. China. He was a visiting scholar at Michigan State University from 2010 to 2011. He was also a research associate professor at CUHK and KAIST in 2012 and 2013, respectively. His research interests include computer graphics, image and video editing, visualization, virtual reality, etc. He has published more than 100 peer-reviewed papers in ACM TOG, IEEE TVCG, IEEE TMM, IEEE TCSVT, ACM TOMM, ACM SIGGRAPH, IEEE VR, ACM MM, IEEE ICME, etc. He is on the editorial board of *Signal, Image and Video Processing* and *Computer Animation and Virtual Worlds*.

Introduction

An image or video is a similar representation and description of the real world with the human visual system, which is among the most important information carriers in human daily activities. An image or video contains many different features of a scene, such as color, texture, illumination, motion, and so on. They are the main sources of information for one to understand the world. In the last decade, with the rapid development of computer technology and the prevalence of the digital cameras and mobile phones, digital image and video editing technology has been paid more and more attention by researchers. Digital image and video editing technology include segmentation [1–4], inpaining [5, 6], color and style transfer [7, 8], colorization [9], decolorization [10], enhancement [11–14], stitching [15, 16], target recognition and tracking [17–21], gamut mapping [22], collage [23], hybrid synthesis [24], line drawing [25–29], cloning [30], visual to sound [31–34], and quality assessment [35–42].

The image and video editing techniques have a wide range of applications in daily life, transportation, military, industry, agriculture, and so on. For example, due to the lack of camera equipment and shooting skills, it is difficult for amateurs to take high-quality images, and there is a growing demand for image editing techniques to change an image to a desired effect. In the field of film and art design, it is often necessary to adjust the color of an image to obtain different artistic styles. In daily social life, one often needs to edit the captured images to get satisfactory results to share on the social media.

Although the existing professional image editing software (e.g., Adobe Photoshop) can complete the complex image editing work, it often requires a series of cumbersome operations to edit the appearance of one image into that of another reference image. Even professional image editors need to spend a lot of time, and it is difficult for non-professionals to use these tools alone to perform image editing. The emergence of image processing technologies such as color transfer not only provides a simple and effective image editing method for image editors, but also makes it possible for non-professional users to edit images according to their own needs.

S. Liu, *Image and Video Color Editing*, Synthesis Lectures on Visual Computing: Computer Graphics, Animation, Computational Photography and Imaging, https://doi.org/10.1007/978-3-031-26030-8_1

Fig. 1.1 An overview of image
and video color editing

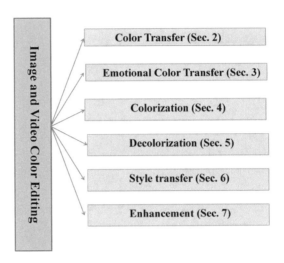

Color is one of the most important features of an image or video. Image and video color
editing aims to change the color appearance of a given image or video, including color transfer, colorization, decolorization, style transfer, etc. This book will overview the progress of
image and video color editing in the last decade. As shown in Fig. 1.1, we summarize the
related techniques, including color transfer, emotional color transfer, colorization, decolorization, style transfer, and enhancement.

1.1   Color Transfer

Color transfer refers to the process of adjusting the color of an image or a video according
to the color of another reference image or video, so that the target image or video has visual
color features similar to the source image or video. For example, one can use the method of
color transfer to convert a red area in the target image into a blue one. After transferring, the
target image or video will show a color style similar to the source image. Another example
is to convert images taken in the morning into night-style images according to the user's
need. Color transfer is often used in image or video processing, image correction, image
enhancement, and other fields.

The early color transfer method was mainly adjusted manually by users. This processing
method needs high time complexity and a large amount of user interaction. In 2001, Reinhard
et al. [43] pioneered a fully automatic image color transfer technique. Given a target image
and a source image, this method can automatically transfer the color of a source image to
the target image, which is simple to implement and highly efficient. In recent years, many
researchers have proposed color transfer methods for images and videos, such as color
transfer based on statistical information (e.g., color mean and variance), and user-assisted

color transfer methods with scribbles, swatches, and so on. Recently, deep learning-based color transfer methods emerged by learning a relationship between the source image and the target image using a deep learning model. As for video color transfer, a direct method is to process a video frame by frame using image transfer techniques. However, temporal inconsistency leads to poor transfer results. By taking into account temporal coherence between neighboring frames, some video color transfer methods also have been developed in recent years.

Image or video color is usually related to the emotion the image or video conveys. It is general to convey emotion using colors. Emotion transfer associates the semantic features of images with emotion. By adjusting the color distribution of the source image or video, the emotional color transfer can convert the emotion of a source image or video into the emotion of a target image or video. The existing emotional color transfer methods include color combination-based methods and deep learning-based methods.

1.2   Colorization and Decolorization

Colorization is a computer-aided process, which aims to give color to a gray image or video. It can be used to enhance black-and-white images, including black-and-white photos, old-fashioned films, and scientific imaging results. Because different colors may produce the same brightness value, colorization is not a problem with deterministic solutions. This technique can promote the efficiency of TV program editing and special effects of film and television production, improve the recognition of biomedical pictures (such as X-ray developing pictures and MRI nuclear magnetic resonance scanning pictures), and enhance the visualization of experimental results in physics and chemistry. It also has a broad application prospect in the camera photography and visualization industry. Early methods require a certain amount of user interaction to produce results that meet the user's needs, but this inevitably makes colorization a time-consuming and laborious process. Later, researchers have developed various automatic image and video colorization methods. Besides natural image colorization, researchers also paid attention to cartoon image colorization, such as in Japanese manga. By exploiting a large set of source images from the Internet or image datasets, deep learning is leveraged to make colorization more efficient by learning a suitable mapping between grayscales and colors.

On the contrary, decolorization is to convert a color image or video into a grayscale one. A grayscale image or video refers to an image or video with only brightness information and without color information. It is the basis of some downstream image processing applications such as pattern recognition, image segmentation, and image enhancement. It is also widely used in daily life. For example, most newspapers and periodicals choose grayscale images as illustrations to save unnecessary expenses. On the other hand, because complete contrast can enhance the main factors, grayscale images have become an artistic choice loved by photography lovers all over the world. In addition, gray images are also widely used in medical photos, surveillance videos, and traffic license plate recognition.

Image or video decolorization is a dimensionality reduction process, i.e., from a 3D matrix to a 1D vector. In this process, information loss will inevitably occur. The key of decolorization is how to make the decolorized image retain one's perception of the target color image as much as possible. The early image decolorization algorithm transforms an input color image from the RGB color space into the $L\alpha\beta$ color space, then obtains the brightness information, and uses the brightness information to represent the gray image. Although this method is simple and easy to implement, it has disadvantages. Especially for areas with the same brightness but different color information, the gray images obtained by the above algorithm are basically the same, so the due information of the image cannot be completely retained. To this end, various image and video decolorization methods have been proposed in the last decade, including the global decolorization method, local decolorization method, and deep learning-based decolorization method.

Different from image decolorization, video decolorization should not only consider the image contrast preservation in each video frame, but also respect the temporal and spatial consistency between video frames. Researchers were devoted to develop decolorization methods by balancing spatial-temporal consistency and algorithm efficiency.

1.3   Style Transfer

Image and video style transfer technology is to transfer the "style" of a source image (e.g., a painting) to a target image or video, so that the target image or video has the same style as the source image. As the extension of image and video color transfer techniques, image and video style transfer methods also have wide applications. For example, it can transform one's photographing or landscape photos into images with the artistic style of *Monet* and *Van Gogh* and other painters.

Before 2015, researchers usually model a specific texture model to represent a style. Because this modeling method is non-parametric, it requires professional researchers to work manually, which is time-consuming and low-efficient. Moreover, each style model is independent of each other, restricting it from practical applications. Gatys et al. [44] began to perform image style transfer by leveraging deep learning networks. In recent years, researchers improved this method and proposed more image style transfers networks to achieve efficient, flexible, and high-quality style results. This book summarized recent progress in image style transfer from the above three aspects.

A video style transfer method transforms a style image (e.g., a van Gogh painting) of a source image into a whole target video sequence. This is a painstaking task if one performs this process manually. Instead, researchers regard video style transfer as an optimization minimization problem or a learning problem. Therefore, some optimization minimization frameworks and deep learning-based frameworks emerged to achieve high-definition video style transfer effects over the past few years.

1.4 Enhancement

In daily life, photography devices may capture videos with a poor quality because of low light-level photographic environment, camera parameter setting, etc. Due to the low visibility, these underexposed images and videos usually fail to present visually pleasing browsing. Image and video enhancement is used to reveal hidden details in underexposed videos and improve video quality, which can contribute to video post-processing.

Until now, researchers have developed various image enhancement methods, including histogram equalization, tone mapping, the Retinex theory-based methods, image fusion, and deep learning-based methods. The histogram equalization [45, 46], transforming the histogram of an original image into a uniformly distributed histogram to increase the image contrast, is a common method for low-exposure image enhancement. By tone mapping, one can adjust the tone curve of an image so as to obtain a more suitable tone curve to enhance a given image or video. The Retinex theory is usually adapted [46–48] for uneven brightness enhancement caused by illumination. The image fusion [49–52] can also be applied for low-exposure image enhancement, which combines the information of multiple images from the same scene to produce a resulting image with optimal details. In recent years, deep learning techniques have been widely used for low-exposure image enhancement, such as [53–60].

Recently, low-exposure video enhancement also attracted much attention (e.g., [61, 62]), due to its wide applications in video processing such as video surveillance and monitoring. A naive way to low-exposure video enhancement is to handle the video frames using the low-exposure image enhancement method. However, it suffers from poor temporal and spatial coherency. To this end, researchers developed methods [11, 61–65] specified to low-exposure video enhancement.

References

1. Y. Zhao, F. Lin, S. Liu, Z. Hu, X. Shi, Y. Bai, C. Shen, Constrained-focal-loss based deep learning for segmentation of spores. IEEE Access. **7**, 165029–165038 (2019)
2. Y. Zhao, S. Liu, Z. Hu, Y. Bai, C. Shen, X. Shi, Separate degree based otsu and signed similarity driven level set for segmenting and counting anthrax spores. Comput. Electron. Agric. **169**, 105230:1-105230:15 (2020)
3. Y. Zhao, S. Liu, Z. Hu, Focal learning on stranger for imbalanced image segmentation. IET Image Process. **16**(5), 1305–1323 (2022)
4. Y. Zhao, S. Liu, Z. Hu, Dynamically blancing class losses in imbalanced deep learning. Electron. Lett. **58**(5), 203–206 (2022)
5. Y. Wei, S. Liu, Domain-based structure-aware image inpainting. Signal Image Video Process. **10**(5), 911–919 (2016)
6. S. Liu, Y. Wei, Fast nearest neighbor searching based on improved VP-tree. Pattern Recognit. Lett. **60–61**, 8–15 (2015)
7. Z. Song, S. Liu, Sufficient image appearance transfer combining color and texture. IEEE Trans. Multimed. **19**(4), 702–711 (2017)

8. S. Liu, T. Zhu, Structure-guided arbitrary style transfer for artistic image and video. IEEE Trans. Multimed. **24**, 1299–1312 (2022)
9. S. Liu, X. Zhang, Automatic grayscale image colorization using histogram regression. Pattern Recognit. Lett. **33**(13), 1673–1681 (2012)
10. S. Liu, H. Wang, X. Zhang, Video decolorization based on CNN and LSTM neural network. ACM Trans. Multimed. Comput. Commun. Appl. **17**(3), 88:1-88:18 (2021)
11. S. Liu, Y. Zhang, Detail-preserving underexposed image enhancement via optimal weighted multi-exposure fusion. IEEE Trans. Consum. Electron. **65**(3), 303–311 (2019)
12. Y. Zhang, S. Liu, Effective underexposed video enhancement via optimal fusion, in *Proceedings of the 9th ACM SIGGRAPH Conference and Exhibition on Computer Graphics and Interactive Techniques in Asia (SIGGRAPH Asia 2016), poster*, Article No. 16 (2016)
13. S. Liu, Y. Zhang, Non-uniform illumination video enhancement based on zone system and fusion, in *Proceedings of International Conference on Pattern Recognition (ICPR)*, pp. 2711–2716 (2018)
14. Y. Zhang, S. Liu, Non-uniform illumination video enhancement based on zone system and fusion. J. Comput.-Aided Des. & Comput. Graph. **29**(12), 2317–2322 (2017)
15. S. Liu, Q. Chai, Shape-optimizing and illumination-smoothing image stitching. IEEE Trans. Multimed. **21**(3), 690–703 (2019)
16. Q. Chai, S. Liu, Shape-optimizing hybrid warping for image stitchings, in *Proceedings of the IEEE International Conference on Multimedia and Expo (ICME)*, Article number: 7552928, pp. 1–6 (2016)
17. Y. Chen, S. Liu, Deep partial occlusion facial expression recognition via improved CNN, in *Proceedings of International Symposium on Visual Computing (ISVC)*, pp. 451–462 (2020)
18. J. Huang, S. Liu, Robust simultaneous localization and mapping in low-light environments. Comput. Animat. Virtual Worlds. **30**(3–4), e1895, 1-10 (2019)
19. S. Liu, Y. Li, G. Hua, Human pose estimation in video via structured space learning and halfway temporal evaluation. IEEE Trans. Circuits Syst. Video Technol. **29**(7), 2029–2038 (2019)
20. S. Liu, G. Hua, Y. Li, 2.5D human pose estimation for shadow puppet animation. KSII Trans. Internet Inf. Syst. **13**(4), 2042–2059 (2019)
21. G. Hua, L. Li, S. Liu, Multi-path affinage stacked-hourglass networks for human pose estimation. Front. Comput. Sci. **14**(4), Article number: 144701 (2020)
22. S. Liu, S. Li, Gamut mapping optimization algorithm based on gamut-mapped image measure (GMIM). Signal Image Video Process. **12**(1), 67–74 (2018)
23. S. Liu, X. Wang, P. Li, J. Noh, Trcollage: efficient image collage using tree-based layer reordering, in *Proceedings of International Conference on Virtual Reality and Visualization (ICVRV)*, pp. 454–455 (2017)
24. S. Liu, J. Wu, Fast patch-based image hybrids synthesis, in *Proceedings of 12th IEEE International Conference on Computer-Aided Design and Computer Graphics (CAD/Graphics)*, pp. 191–197 (2011)
25. J. Zhang, Z. Liu, S. Liu, Computer simulation of archaeological drawings. J. Cult. Herit. **37**, 181–191 (2019)
26. S. Liu, Z. Liu, Style enhanced line drawings based on multi-feature, Multimed. Tools Appl. **81**, (2022)
27. J. Zhang, S. Liu, J. Fan, J. Huang, M. Pei, NK-CDS: a creative design system for museum art derivatives. IEEE Access. **8**(1), 29259–29269 (2020)
28. K. Fan, S. Liu, A caricature-style generating method for portrait photo. J. Comput.-Aided Des. & Comput. Graph. **34**(1), 1–8 (2022)

29. K. Fan, S. Liu, W. Lu, SemiPainter: Learning to draw semi-realistic paintings from the manga line drawings and flat shadow, in *Proceedings of Computer Graphics International (CGI)*, (Geneva, Switzerland, 2022)

30. H. Cheng, K. Wang, S. Liu, Adaptive color-style-aware image cloning. J. Graph. **38**(5), 700–705 (2017)

31. S. Liu, S. Li, H. Cheng, Towards an end-to-end visual-to-raw-audio generation with GANs. IEEE Trans. Circuits Syst. Video Technol. **32**(3), 1299–1312 (2022)

32. S. Li, S. Liu, D. Manocha, Binaural audio generation via multi-task learning. ACM Trans. Graph. **40**(6), 243:1–243:13 (2021)

33. J. Hao, S. Liu, Q. Xu, Controlling eye blink for talking face generation via eye conversion, in *Proceeding of SIGGRAPH Asia Technical Communications, Article 2021*, vol. 1, pp. 1–4 (2021)

34. S. Liu, J. Hao, Generating talking face with controllable eye movements by disentangled blinking feature. IEEE Trans. Vis. Comput. Graph. **28**, (2022)

35. C. Zhang, Z. Huang, S. Liu, J. Xiao, Dual-channel multi-task CNN for no-reference screen content image quality assessment. IEEE Trans. Circuits Syst. Video Technol. **32**, (2022)

36. R. Gao, Z. Huang, S. Liu, QL-IQA: Learning distance distribution from quality levels for blind image quality assessment. Signal Process. Image Commun. **101**(116576), 1–11 (2022)

37. Z. Huang, S. Liu, Perceptual hashing with visual content understanding for reduced-reference screen content image quality assessment. IEEE Trans. Circuits Syst. Video Technol. **31**(7), 2808–2823 (2021)

38. Z. Huang, S. Liu, Perceptual image hashing with texture and invariant vector distance for copy detection. IEEE Trans. Multimed. **23**, 1516–1529 (2021)

39. R. Gao, Z. Huang, S. Liu, Multi-task deep learning for no-reference screen content image quality assessment, in *Proceedings of the 27th International Conference on Multimedia Modeling (MMM)*, (Prague, Czech Republic, 2021), pp. 213–226

40. S. Liu, Z. Huang, Efficient image hashing with invariant vector distance for copy detection. ACM Trans. Multimed. Comput. Commun. Appl. **15**(4), 106:1–106:22 (2019)

41. Z. Huang, S. Liu, Robustness and discrimination oriented hashing combining texture and invariant vector distance, in *Proceedings of ACM Multimedia*, pp. 1389–1397 (2018)

42. C. Zhang, S. Liu, No-reference omnidirectional image quality assessment based on joint network, in *Proceedings of the 30th ACM International Conference on Multimedia (ACM MM)*, (Lisbon, Portugal, 2022), pp. 943–951

43. E. Reinhard, M. Ashikhmin, B. Gooch, P. Shirley, Color transfer between images. IEEE Comput. Graph. & Appl. **21**(5), 34–41 (2001)

44. L.A. Gatys, A.S. Ecker, M. Bethge, Image style transfer using convolutional neural networks, in *Proceedings of IEEE Conference on Computer Vision and Pattern Recognition (CVPR)*, pp. 2414–242 (2016)

45. A.W.M. Abdullah, M.H. Kabir, M.A.A. Dewan et al., A dynamic histogram equalization for image contrast enhancement. IEEE Trans. Consum. Electron. **53**(2), 593–600 (2007)

46. H. Malm, M. Oskarsson, E. Warrant, et al., Adaptive enhancement and noise reduction in very low light-level video, in *Proceedings of IEEE International Conference on Computer Vision*, pp. 1–8 (2007)

47. B. Li, S. Wang, Y. Geng, Image enhancement based on retinex and lightness decomposition, in *Proceedings of IEEE International Conference on Image Processing (ICIP)*, pp. 3417–3420 (2011)

48. J.H. Jang, S.D. Kim, J.B. Ra, Enhancement of optical remote sensing images by subband-decomposed multiscale Retinex with hybrid intensity transfer function. IEEE Geosci. & Remote. Sens. Lett. **8**(5), 983–987 (2011)

49. I. Adrian, R. Ramesh, J.Y. Yu, Gradient domain context enhancement for fixed cameras. Int. J. Pattern Recognit. & Artif. Intell. **19**(4), 533–549 (2011)

50. R. Raskar, A. Ilie, J. Yu, Image fusion for context enhancement and video surrealism, in *Proceedings of ACM SIGGRAPH Courses*, pp. 85–152 (2004)

51. Y. Cai, K. Huang, T. Tan, et al., Context enhancement of nighttime surveillance by image fusion, in *Proceedings of International Conference on Pattern Recognition*, pp. 980–983 (2006)

52. X. Gao, S. Liu, DAFuse: a fusion for infrared and visible images based on generative adversarial network. J. Electron. Imaging **31**(4), 1–18 (2022)

53. K.G. Lore, A. Akintayo, S. Sarkar, LLNet: a deep autoencoder approach to natural low-light image enhancement. Pattern Recognit. **61**, 650–662 (2017)

54. F. Lv, F. Lu, J. Wu, C. Lim, MBLLEN: Low-light "image/video enhancement using CNNs," in *Proceedings of British Machine Vision Conference (BMVC)*, (Newcastle, UK, 2018)

55. M. Zhu, P. Pan, W. Chen, Y. Yang, EEMEFN: Low-light image enhancement via edge-enhanced multi-exposure fusion network, in *Proceedings of the AAAI Conference on Artificial Intelligence*, pp. 13106–13113 (2020)

56. Y. Jiang, X. Gong, D. Liu, Y. Cheng, C. Fang, X. Shen, J. Yang, P. Zhou, Z. Wang, EnlightenGAN: Deep light enhancement without paired supervision. IEEE Trans. Image Process. **30**, 2340–2349 (2021)

57. W. Ren, S. Liu, L. Ma et al., Low-light image enhancement via a deep hybrid network. IEEE Trans. Image Process. **28**(9), 4364–4375 (2019)

58. R. Yu, W. Liu, Y. Zhang, Z. Qu, D. Zhao, B. Zhang, DeepExposure: Learning to expose photos with asynchronously reinforced adversarial learning, in *Proceedings of Advances in Neural Information Processing Systems*, (2018)

59. C. Chen, Q. Chen, J. Xu, V. Koltun, Learning to see in the dark, in *Proceedings of IEEE/CVF Conference on Computer Vision and Pattern Recognition (CVPR)*, pp. 3291–3300 (2018)

60. R. Wang, Q. Zhang, C.-W. Fu, X. Shen, W.-S. Zheng, J. Jia, Underexposed photo enhancement using deep illumination estimation, in *Proceedings of IEEE/CVF Conference on Computer Vision and Pattern Recognition (CVPR)*, pp. 6842–6850 (2019)

61. Q. Zhang, Y. Nie, L. Zhang et al., Underexposed video enhancement via perception-driven progressive fusion. IEEE Trans. Vis. & Comput. Graph. **22**(6), 1773–1785 (2016)

62. X. Dong, L. Yuan, W. Li, et al., Temporally consistent region-based video exposure correction, in *Proceedings of IEEE International Conference on Multimedia and Expo*, pp. 1–6 (2015)

63. C. Chen, Q. Chen, M. Do, V. Koltun, Seeing motion in the dark, in *Proceedings of IEEE/CVF International Conference on Computer Vision (ICCV)*, pp. 3184–3193 (2019)

64. C. Zheng, Z. Li, Y. Yang, S. Wu, Single image brightening via multi-scale exposure fusion with hybrid learning. IEEE Trans. Circuits Syst. Video Technol. **31**(4), 1425–1435 (2020)

65. D. Triantafyllidou, S. Moran, S. McDonagh, S. Parisot, G. Slabaugh, Low light video enhancement using synthetic data produced with an intermediate domain mapping, in *Proceedings of European Conference on Computer Vision (ECCV)*, pp. 103–119 (2020)

Color Transfer

Image or video appearance features (e.g., color, texture, tone, illumination, and so on) reflect one's visual perception and direct impression of an image or video. Given a source image (video) and a target image (video), the image (video) color transfer technique aims to process the color of the source image or video (note that the source image or video is also referred to as the reference image or video in some literature) to make it look like that of the target image or video, i.e., transferring the appearance of the target image or video to that of the source image or video, which can thereby change one's perception of the source image or video. This chapter will introduce the techniques about image and video color transfer.

2.1   Image Color Transfer

Color is one of the most important visual information of an image. Color transfer, a widely used technique in digital image processing, refers to transferring the color distribution of a source image to another target image, so that the edited target image can have a color distribution similar to the source image. The existing color transfer methods can be classified into three categories, namely color transfer methods based on statistical information [1], color transfer methods based on geometry [2], and color transfer methods based on user interaction [3].

2.1.1   Color Transfer Based on Statistical Information

Global Color Transfer. In 2001, Reinhard et al. [1] pioneered the technique of image color transfer between images. Given a target image and a source image, with two statistical indexes (i.e., mean and standard deviation) of an image, this method simply and efficiently transfers the global color statistical information from the target image to the source image,

© The Author(s), under exclusive license to Springer Nature Switzerland AG 2023   9
S. Liu, *Image and Video Color Editing*, Synthesis Lectures on Visual Computing: Computer
Graphics, Animation, Computational Photography and Imaging,
https://doi.org/10.1007/978_3_031_26030_8_2

so that the source image has a color distribution similar to the target image, that is, the two images have a similar visual appearance. This method is performed in the $l\alpha\beta$ color space due to the independence between three color channels as follows:

$$l' = \frac{\sigma_t^l}{\sigma_s^l}(l - \langle l \rangle),$$

$$\alpha' = \frac{\sigma_t^\alpha}{\sigma_s^\alpha}(\alpha - \langle \alpha \rangle), \qquad (2.1)$$

$$\beta' = \frac{\sigma_t^\beta}{\sigma_s^\beta}(\beta - \langle \beta \rangle),$$

where l, α, and β are the color components of each pixel in the target image, and l', α', and β' the color components of each pixel in the resulting image, respectively. The $\langle l \rangle$, $\langle \alpha \rangle$, and $\langle \beta \rangle$ are the mean values of l, α, and β. The σ_s^l, σ_s^α, and σ_s^β are the standard deviation of the source image and σ_t^l, σ_t^α, and σ_t^β are the standard deviation of the target image, respectively. This algorithm needs to calculate the mean and standard deviation of all three channels of the two images, in which the mean is used to represent the color distribution characteristics of the whole image, and the local details of the image are represented by the standard deviation. As this method treats all pixels in a global manner, it belongs to *global* color transfer techniques.

Following Reinhard et al.'s idea, many impressive image color transfer methods emerged. Piti et al. [4] proposed a new global color transfer method, which can migrate the color probability density function of any target image to another source image. Pouli and Reinhard [5] proposed a *progressive* color transfer approach to the histograms for a series of consecutive scales. This method can robustly transfer the color palette between images, allowing one to control the amount of color matching in an intuitive manner. However, when there is a large difference in color distribution between the source image and the target image, the effect of the above global color transfer method may be less satisfactory. To this end, *local* color transfer methods were developed.

Local Color Transfer. Tai et al. [6] used the Gaussian mixture model to realize local color transfer between images. In this method, the expectation-maximization method is exploited to improve the accuracy of color cluster distribution. Later, Irony et al. [7] proposed a local color transfer method based on higher-level features. This method transfers the color information to each pixel of the target gray image by using a segmented reference image. Nguyen et al. [8] proposed a color transfer method considering color gamut, which can control the color gamut of the transfer result within the range of the color gamut of the source image, making it more consistent with the color distribution of the source image. Hwang et al. [9] presented a color transfer method based on moving least squares. This method can produce robust results for different images caused by different camera parameters, different shooting times, or different lighting conditions. However, because a large number of feature points are needed as control points, this method requires that the target image and source image share the same scene, which limits its practical applications. Liu et al. [10] presented a selective color transfer method between images via an ellipsoid color mixture map. In this method, ellipsoid hulls are employed to represent the color statistics of the images. This method computes the ellipsoid hulls of the source and target images and produces a color mixture map to determine the blending weight of pixels in the output image, according to the color and distance information instead of using image segmentation. By mixing the images using the color mixture map, high-quality local color transfer results can be generated (See Fig. 2.1).

The method of color transfer can be carried out not only between two color images, but also between color images and gray images. The latter is also called gray image colorization. We will discuss this case in the next Chap. 3.

Although the color transfer technique can successfully change the appearance of an image (e.g., changing an image from sunny to cloudy), these methods cannot generate new contents

(a) source image   (b) target image

(c) color mixture map   (d) final result

Fig. 2.1 The selective color transferring result of a flower scene [10] (e.g., new leaves on a dead tree) in the image. When users need to create new contents in the target image, the color transfer method may fail.

2.1.2   Color Transfer Based on User Interaction

In the process of digital image processing, professional software such as Photoshop can help users edit the image appearance flexibly and effectively (e.g., adjusting color, brightness, fusion texture, etc.). However, non-professional users lack the skills to operate such softwares. For the convenience of users, stroke-based techniques have been introduced into many image editing fields (e.g., color transfer, image segmentation, etc.).

Luan et al. [11] used a brush-based interactive system for image color transfer. This method needs to mark the pixels of all the areas for color transfer with a brush in the image. Wen et al. [12] leveraged stroke-based editing and propagation to develop an interactive system for image color transfer. In comparison with Luan et al.'s approach [11], with this method users can get ideal results through a smaller amount of stroke inputs. However, this method cannot deal with the transfer of discrete unconnected regions.

In recent years, stroke-based editing propagation technique has been widely studied. The information that needs to be edited by the user will be automatically propagated to other similar image editing areas with the stroke. To the best of our knowledge, Levin et al. [13] proposed the first editing propagation framework, which can complete more accurate gray image colorization free of any image segmentation or region tracking. Editing propagation follows a simple criterion: pixels that are spatially adjacent and have similar attributes should have similar colors after editing. Therefore, the color transfer problem is formulated as a global optimization problem. The strokes with editing parameters are used to mark the corresponding regions on the gray image. Based on the above criteria, the marking operation on the strokes is propagated to the adjacent similar regions in the image. The method of editing propagation provides users with a simple and efficient way to edit the color of an image. Later, some researchers used global editing propagation techniques to propagate the editing operation on the stroke to the similar areas in the whole image.

In the above editing propagation methods, the editing information on the strokes needs to be set through parameters. However, in many cases, it is difficult for users to tweak these parameters to meet a desired editing goal. To this end, An et al. [3] introduced the reference image in the editing propagation to provide general editing information, so that users can directly obtain the editing parameters of the image by marking the stroke on the region of interest of the color reference image and then use the editing propagation method to spread the required editing to the whole target image.

However, the existing editing propagation methods assume that the user's stroke marking is accurate, i.e., the marked stroke only covers the pixels of the user's region of interest. When the stroke mark does not meet this assumption, it will directly affect the final editing and propagation results. Therefore, Subr et al. [14] and Sener et al. [15] used Markov random fields to solve this problem. However, the introduction of the Markov random fields leads to a high computational complexity, which limits its practical application.

This book focuses on color transfer between images. However, except for color, more features (e.g., texture, contrast, and tone) contribute to the appearance of an image. More generally, given a relationship between two images, it can be replicated on a new image by the technique of image analogies [16].

2.2   Hybrid Transfer

Image Texture Transfer. Texture is another critical image feature. In recent years, the texture synthesis technique, especially the sample-based texture synthesis method, has made remarkable progress [17]. The core idea of sample-based texture synthesis method is to generate a new texture similar to the sample according to the surface features of a given texture sample. Image analogy [16], also known as texture transfer [18], refers to synthesizing new textures in a target image to make the target image and the reference image share a similar texture appearance. Bonneel et al. [19] used texture transfer to enhance the appearance of 3D models. With a photograph as the source image, a series of computer-synthesized target images become more realistic by using color and texture transfer to improve the roughness of the appearance of the target images [20]. Zhang and Liu [21] addressed texture transfer in the frequency domain. This algorithm is efficient due to various FFT (Fast Fourier Transform) algorithms for transforming images to the frequency domain, which can utilize any of the texture synthesis algorithms to generate a texture of the proper size and then transfer this texture to the target image. A set of tools and parameters are also provided to control the process of this method. Diamanti et al. [22] manually annotated the sample image to mark the areas that need to generate texture in the process of texture synthesis. Although texture transfer can synthesize new content in a target image, it often covers the structural details that should be retained in the target image.

Shih et al. [23] used the image of a scene at one time to predict the appearance of the scene at another time. This method can also realize the prediction of the corresponding night image by taking the daytime image as the source image. However, if the night image is the source image, because only color transfer is considered, the details of the dark places at night will be lost, and it then fails to predict the corresponding daytime image.

Image Tone Transfer. The pixel colors may vary greatly in one image due to different lighting environments, leading to distinct visual discontinuities. To this end, Liu et al. [24] proposed a multi-toning image adjustment method via color transferring. An unsupervised method is used to cluster the source and the target image based on color statistics. A correspondence is established via matching the texture and luminance between the subsets in the source and target images. The color is then transferred from the matched pixels in the source to the target. Finally, the graph cut is utilized to optimize the seams between different subsets in the target image.

Bae et al. [25] proposed to transfer the tone from a source photo to make a target image have a photo graphics look. Based on a two-scale nonlinear decomposition, this method modifies different layers of an image via their histograms. A Poisson correction mechanism is designed to prevent potential gradient reversal and preserve image detail.

Without a source image, the tone of a given can also be creatively adjusted by a user. For example, Lischinski et al. [26] proposed to adjust the tone of an image in the regions of interest with strokes. To achieve a finer tone control, Farbman et al. [27] manipulated the tone of an image from the coarsest to the finest level of detail with a multi-scale decomposition of the image.

Tone reproduction for dynamic range reduction also gains much interest from researchers. Various operators (e.g., logarithms, histograms, sigmoids, etc.) have been devised for tone curves [28–33].

The above color transfer and texture transfer methods can process the image appearance from the perspective of a single feature, i.e., color or texture, so that the target image is similar to the source image in terms of color or texture. However, the color and texture of an image cannot be changed at the same time by using color transfer or texture transfer alone.

Image Hybrid Transfer. In order to tackle this challenge, Okura et al. [34] first proposed an image hybrid transfer method by combining color transfer and texture transfer. This method can take full advantage of the two transfer methods. It can not only change the color of the target image, but also can generate new content in the target image. This method uses a ternary image group as input: a target image, a sample image, and a source image. The target image and sample image are required to be photos of the *same* scene taken at different times and need to have the same perspective. The source image needs to have some visual similarity with the target image. By predicting the probability of a successful color transfer between the sample image and the source image, this method calculates the error between the target image and the sample image, so as to judge the areas requiring color transfer and texture transfer in the target image. Among them, the region with small errors is processed by color transfer. For regions with large errors, texture transfer is required. This method primarily uses the fact that the sample image is the real transfer result of the target image to predict and process the target image and finally make the target image have the appearance similar to the sample image. However, this method suffers from two limitations. On one hand, the requirements for the input image are relatively strict, which limits its practical application to a great extent; on the other hand, limited by the methods based on color transfer, texture transfer, and image matching, inaccurate matching and large errors may occur in the results of color transfer and texture transfer and therefore fail to generate an ideal image transfer result.

Song and Liu [35] presented a novel image hybrid transfer method combining color and texture features. The algorithm does not require the user to provide a ternary image group. Instead, it mainly relies on feature extraction and feature matching to automatically judge the regions that need texture transfer or color transfer in the source and target images. This method allows users to mark the areas to be transferred using strokes. Color transfer and

Table 2.1 Image color transfer methods

Methods	References
Global color transfer	Reinhard et al. [1], Pitié et al. [4]
	Pouli and Reinhard [5]
Local color transfer	Irony et al. [7], Tai et al. [6]
	Liu et al. [10], Hwang et al. [9]
	Nguyen et al. [8]
User-assisted color transfer	Levin et al. [13], Luan et al. [11]
	Wen et al. [12], An et al. [3]
	Sener et al. [15], Subr et al. [14]
Hybrid transfer	Okura et al. [34], Song and Liu [35]
	Liu and Song [36]
Deep learning-based	Lee and Lee [38], He et al. [39]
Color transfer	Liao et al. [40], Liu et al. [37]

texture transfer are then performed in the transferring task between corresponding regions. Later, Liu and Song [36] extended the above method to deal with image appearance with multi-source images.

Albeit successful, the hybrid transfer methods may fail if there are large illumination differences between the source image and the target image. Especially when there is strong illumination in the image, the color deviation caused by illumination or the light that can be seen in the target image will affect the result of color transfer. Because the traditional color transfer method does not consider the problems caused by these lighting factors, it cannot produce results similar to the source image. To this end, Liu et al. [37] devised an appearance transfer method between images with complex illumination images via deep learning. We will elaborate on this method in the next section.

Table 2.1 summarizes the recent image color transfer methods.

2.3 Video Color Transfer

Either image or image sequence can arouse emotion. In 2004, Wang and Huang [41] proposed an image sequence color transfer algorithm, that can render an image sequence with color distribution transferred from three source images. The mean and variance of the colors are computed in each source image, which are then interpolated with a color variation curve (CVC) to generate in-between color transfer results. However, this method suffers from a limitation in that only three source images can be allowed for interpolation for an in-between image sequence.

Wang et al. [42] further presented a more general image sequence color transfer method. Given a video clip as the input source, by analyzing the color mood variation with B-spline curves, they create a resulting image sequence from only one image. Instead of only supporting a linear or parabolic color variation curve, this method employs a generalized color variation curve (GCVC) for more flexible control for color transfer across in-between images. The B-spline allows the user to more effectively design a curve by the control points, which can achieve similar results to the brute force method.

In movie productions, artists usually painstakingly tweak the color palette of a film to achieve a desired appearance, which is also known as *color grading*. Bonneel et al. [43] proposed an example-based video color grading approach, that automatically transfers the color palette of a source video clip to a target video. Firstly, a per-frame color transform is computed to map the color distribution from the target video to that of the source video. To reduce artifacts (e.g., bleeding and flickering), a differential geometry-based scheme is then devised to interpolate the transformations so that their curvature can be minimized. Finally, the keyframes that best represent this interpolated transformation curve are determined for the subsequent process of color grading.

Aiming at a full nonlinear and non-parametric color mapping in the 3D color space, Hwang et al. [9] introduced moving least squares into the color transfer and proposed a scattered point interpolation framework with a probabilistic modeling of the color transfer in the 3D color space to achieve video color transfer.

Recently, Liu and Zhang [44] proposed a novel temporal-consistency-aware video color transfer method via a quaternion distance metric. They extracted keyframes from the target video and transfer their color from the source image by the Gaussian Mixture Models (GMM) based on a soft segmentation algorithm. By matching the pixel through a quaternion-based distance metric, they transfer color from keyframes to the in-between frames in an iterative manner. An abnormal color correction mechanism is further devised to improve the resulting color quality. This method is capable of better preserving the temporal consistency between video frames and alleviating the color artifacts than the state-of-the-art methods (Fig. 2.2).

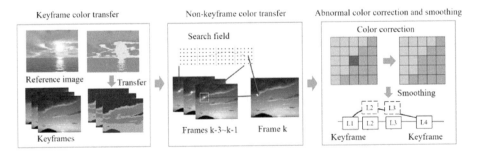

Fig. 2.2 The framework of the temporal-consistency-aware video color transfer method [44]

2.4   Deep Learning-Based Color Transfer

Deep learning-based methods aim to transform the transfer problem into a nonlinear regression problem and then solve the regression problem through the powerful ability of the deep learning method, so as to obtain an appearance mapping relationship between the source image and the target image. With the trained network model, users can apply the learned specific mapping relationship to an input target image to obtain a similar appearance transfer result.

In addition to the breakthrough progress in handwritten digit recognition, convolutional neural network (CNN) has also demonstrated excellent capabilities in other computer vision applications, such as digital image classification [45], pedestrian detection [46, 47], image super-resolution [48], and image enhancement [49]. Especially in the field of digital image classification, the introduction of deep neural networks greatly improves the accuracy of image classification. In recent years, in the field of image color transfer, the introduction of large-scale data and deep learning methods is also a research hotspot. In the two application fields of grayscale image coloring and image artistic style transfer [50], researchers took the lead in implementing the image appearance transfer method based on depth neural network technology.

He et al. [39] proposed a neural color transfer between images. They use neural representations (e.g., convolutional features of pre-trained VGG-19) to achieve dense semantically meaningful matching between the source and target images. This method is free of undesired distortions in edges or detailed patterns by performing local color transfer in the image domain instead of the feature space.

Liao et al. [40] proposed "deep image analogy" to achieve visual attribute transfer (e.g., color, texture, tone, and style) across images with very different appearances but perceptually similar semantic structures. Aiming at better representations for semantic structures, a deep Convolutional Neutral Network (CNN) [45] was employed to construct a feature space in which to form image analogies.

Lee and Lee [38] transferred the appearance of noon images to night images using CNN. Liu et al. [37] designed a deep transfer framework for images with complex illumination (see Fig. 2.3). This method consists of illumination transfer and color transfer. CNN is employed to extract the hierarchical feature maps, of which the former is carried out in the illumination channel, while for the other two channels, the latter is performed to transfer the color distribution to the target image. This method can achieve progressive transfer using the histogram reshaping method with hierarchical feature maps extracted from the CNN. The final appearance transfer results are output by combining the illumination transfer and color transfer. Finally, the joint bilateral filter is adopted to smooth the noises so as to further optimize the transfer results.

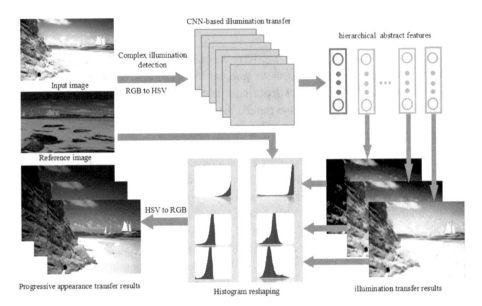

Fig. 2.3 The framework of the progressive complex illumination image appearance transfer method with CNN [37]

References

1. E. Reinhard, M. Ashikhmin, B. Gooch, P. Shirley, Color transfer between images. IEEE Comput. Graph. Appl. **21**(5), 34–41 (2001)
2. Y. Hacohen, E. Shechtman, D.B Goldman et al., Non-rigid dense correspondence with applications for image enhancement. ACM Trans. Graph. **30**(4), 76–79 (2011)
3. X. An, F. Pellacini, User-controllable color transfer. Comput. Graph. Forum **29**(2), 263–271 (2010)
4. F. Pitié, A.C. Kokaram, R. Dahyot, Automated colour grading using colour distribution transfer. Comput. Vis. Image Underst. **107**(1–2), 123–137 (2007)
5. T. Pouli, E. Reinhard, Progressive color transfer for images of arbitrary dynamic range. Comput. Graph. **35**, 67–80 (2011)
6. Y.W. Tai, J. Jia, C.K. Tang, Soft color segmentation and its applications. IEEE Trans. Pattern Anal. Mach. Intell. **29**(9), 1520–1537 (2007)
7. R. Irony, D. Cohen-Or, D. Lischinski, Colorization by example, in *Proceedings of Eurographics Symposium on Rendering Techniques*, pp. 201–210 (2005)
8. R.M.H. Nguyen, S.J. Kim, M.S. Brown, Illuminant aware gamut-based color transfer. Comput. Graph. Forum **33**(7), 319–328 (2014)
9. Y. Hwang, J.Y. Lee, I.S. Kweon, et al., Color transfer using probabilistic moving least squares, in *Proceedings of IEEE Conference on Computer Vision and Pattern Recognition*, pp. 3342–3349 (2014)
10. S. Liu, H. Sun, X. Zhang, Selective color transferring via ellipsoid color mixture map. J. Vis. Commun. Image Represent. **23**(1), 173–181 (2012)

11. Q. Luan, F. Wen, Y. Xu, Color transfer brush, in *Proceedings of the 15th Pactific Conference on Computer Graphics and Applications*, pp. 465–468 (2007)
12. C. Wen, H. Chang-Hsi, B. Chen B, et al., Example-based multiple local color transfer by strokes. Comput. Graph. Forum **27**(7), 1765–1772 (2008)
13. A. Levin, D. Lischinski, Y. Weiss, Colorization using optimization. ACM Trans. Graph. **23**(3), 689–694 (2004)
14. K. Subr, S. Paris, C. Soler, J. Kautz J, Accurate binary image selection from inaccurate user input. Comput. Graph. Forum **32**(2), 41–50 (2013)
15. O. Sener, K. Ugur, A.A. Alatan, Error-tolerant interactive image segmentation using dynamic and iterated graph-cuts, in *Proceedings of the 2nd ACM International Workshop on Interactive Multimedia on Mobile and Portable Devices*, pp. 9–16 (2012)
16. A. Hertzmann, C.E. Jacobs, N. Oliver, et al., Image analogies, in *Proceedings of ACM SIGGRAPH*, pp. 327–340 (2001)
17. L. Wei, S. Lefebvre, V. Kwatra, G. Turk, State of the art in example-based texture synthesis, in *Proceedings of Eurographics 2009, State of the Art Report, EG-STAR*, pp. 1–25 (2009)
18. A.A. Efros, W.T. Freeman, Image quilting for texture synthesis and transfer, in *Proceedings of ACM SIGGRAPH*, pp. 341–346 (2015)
19. N. Bonneel, M.V.D. Panne, S. Lefebvre et al., Proxy-guided texture synthesis for rendering natural scenes, in *Proceedings of Vision, Modeling & Visualization Workshop*, pp. 87–95 (2010)
20. M.K. Johnson, K. Dale, S. Avidan S, et al., CG2Real: Improving the realism of computer generated images using a large collection of photographs. IEEE Trans. Vis. Comput. Graph. **17**(9), 1273–1285 (2010)
21. X. Zhang, S. Liu, Texture transfer in frequency domain, in *Proceedings of the International Conference on Image and Graphics (ICIG)*, pp. 123–128 (2011)
22. O. Diamanti, C. Barnes, S. Paris, E. Shechtman, O. Sorkine-Hornung, Synthesis of complex image appearance from limited exemplars. ACM Trans. Graph. **34**(2), 22:1–22:14 (2015)
23. Y. Shih, S. Paris, F. Durand, W.T. Freeman, Data-driven hallucination of different times of day from a single outdoor photo. ACM Trans. Graph. **32**(6), 2504–2507 (2013)
24. S. Liu, X. Wang, Q. Peng, Multi-toning image adjustment. Comput. Aided Draft. Des. Manuf. **21**(2), 62–72 (2011)
25. S. Bae, S. Paris, F. Durand, Two-scale tone management for photo graphic look. ACM Trans. Graph. **25**(3), 637–645 (2006)
26. D. Lischinski, Z. Farbman, M. Uyttendaele, Interactive local adjustment of tonal values. ACM Trans. Graph. **25**(3), 646–653 (2006)
27. Z. Farbman, R. Fattal, D. Lischinski, R. Szeliski, Edge-preserving decompositions for multi-scale tone and detail manipulation. ACM Trans. Graph. **27**(3), 67:1–67:10 (2008)
28. F. Drago, K. Myszkowski, T. Annen, N. Chiba, Adaptive logarithmic mapping for displaying high contrast scenes. Comput. Graph. Forum **22**(3), 419–426 (2003)
29. G. Ward, H. Rushmeier, C. Piatko, A visibility matching tone reproduction operator for high dynamic range scenes. IEEE Trans. Vis. Comput. Graph. **3**(4), 291–306 (1997)
30. E. Reinhard, M. Stark, P. Shirley, J. Ferwerda, Photographic tone reproduction for digital images. ACM Trans. Graph. **21**(3), 267–276 (2002)
31. F. Durand, J. Dorsey, Fast bilateral filtering for the display of high-dynamic-range images. ACM Trans. Graph. **21**(3), 257–266 (2002)
32. R. Fattal, D. Lischinski, M. Werman, Gradient domain high dynamic range compression. ACM Trans. Graph. **21**(3), 249–256 (2002)
33. Y. Li, L. Sharan, E. Adelson, Compressing and companding high dynamic range images with subband architectures. ACM Trans. Graph. **24**(3), 836–844 (2005)

34. F. Okura, K. Vanhoey, A. Bousseau, A.A. Efros, G. Drettakis, Unifying color and texture transfer for predictive appearance manipulation. Comput. Graph. Forum **34**(4), 53–63 (2015)
35. Z. Song, S. Liu, Sufficient image appearance transfer combining color and texture. IEEE Trans. Multimed. **19**(4), 702–711 (2017)
36. S. Liu, Z. Song, Multi-source image appearance transfer based on edit propagation. J. Zhengzhou Univ. (Engineering Science) **39**(5), 22–27 (2018)
37. S. Liu, Z. Song, X. Zhang, T. Zhu, Progressive complex illumination image appearance transfer based on CNN. J. Vis. Commun. Image Represent. **64**, 102636:1–102636:11 (2019)
38. J. Lee, S. Lee, Hallucination from noon to night images using CNN, in *Proceedings of SIGGRAPH Asia Posters*, Article no. 15 (2016)
39. M. He, J. Liao, L. Yuan, P.V. Sander, Neural color transfer between images, in *Proceedings of IEEE Conference on Computer Vision and Pattern Recognition*, pp. 1–14 (2017)
40. J. Liao, Y. Yao, L. Yuan, G. Hua, S.B. Kang, Visual attribute transfer through deep image analogy (2017). arXiv:1705.01088
41. C.M. Wang, Y.H. Huang, A novel color transfer algorithm for image sequences. J. Inf. Sci. Eng. **20**(6), 1039–1056 (2004)
42. C.M. Wang, Y.H. Huang, M.L. Huang, An effective algorithm for image sequence color transfer. Math. Comput. Model. **44**(7), 608–627 (2006)
43. N. Bonneel, K. Sunkavalli, S. Paris, H. Pfister, Example-based video color grading. ACM Trans. Graph. **32**(4), 39:1–39:12 (2013)
44. S. Liu, Y. Zhang, Temporal-consistency-aware video color transfer, in *Proceedings of Computer Graphics International (CGI)*, pp. 464–476 (2021)
45. A. Krizhevsky, I. Sutskever, G.E. Hinton, ImageNET classification with deep convolutional neural networks. Commun. ACM **60**(6), 84–90 (2017)
46. W. Ouyang, X. Wang, Joint deep learning for pedestrian detection, in *Proceedings of IEEE International Conference on Computer Vision (ICCV)*, pp. 2056–2063 (2013)
47. S. Zheng, A. Yuille, Z. Tu, Detecting object boundaries using low-, mid-, and high-level information. Comput. Vis. Image Underst. **114**(10), 105–1067 (2010)
48. C. Dong, C.L. Chen, K. He, et al., Learning a deep convolutional network for image super-resolution, in *Proceedings of European Conference on Computer Vision (ECCV)*, pp. 184–199 (2014)
49. Z. Yan, H. Zhang, B. Wang, S. Paris, Y. Yu, Automatic photo adjustment using deep learning. ACM Trans. Graph. **35**(2), 11:1–11:15 (2016)
50. L.A. Gatys, A.S. Ecker, M. Bethge, A neural algorithm of artistic style, in *Proceedings of IEEE Computer Vision and Pattern Recognition (CVPR)* (2015)
22=

Emotional Color Transfer

Color is widely used to represent the emotion of an image. The emotion of a target image can be changed by adjusting its color distribution. Emotional color transfer aims to transfer the emotion of the source image to that of the target image by changing the color distribution of the target image. Emotional transfer can be widely used in multimedia processing, advertising, and movie special effects. For example, through the technique of emotional color transfer, one can transfer a given image into a brighter and fresher one with a happy emotion. This technique has great potential to inspire designers and improve their work efficiency. The existing emotion transfer methods can be classified into two categories, emotional color transfer method based on color combination and emotional color transfer method based on deep learning. Table 3.1 summarizes the emotional color transfer methods in recent years.

3.1 Color Combination-Based Emotional Color Transfer

The emotional color transfer method based on color combination mainly considers the low-level semantic information, i.e., color of an image, and then transfers the color distribution of the image to the target image, so that the color distribution of the target image is consistent with that of the source image.

3.1.1 Emotion Transfer Based on Histogram

Yang and Peng [1] considered the relationship between image emotion and color in color transfer and used histogram matching to maintain spatial consistency between image colors. This method uses the RGB color space for image classification and color transformation. In order to align the target image more closely with the source image, Xiao and Ma's local color transfer method [9] is employed to match the color distribution of the target image

© The Author(s), under exclusive license to Springer Nature Switzerland AG 2023
S. Liu, *Image and Video Color Editing*, Synthesis Lectures on Visual Computing: Computer Graphics, Animation, Computational Photography and Imaging,
https://doi.org/10.1007/978-3-031-26030-8_3

Table 3.1 Emotional color transfer methods

Methods	References
Emotional color transfer based on histogram	Yang and Peng [1]
Emotional color transfer based on emotion word	Wang et al. [2]
Emotional color transfer based on color combination	He et al. [3], Liu and Pei [4]
Deep learning-based emotional color transfer	Peng et al. [5], Liu et al. [6]
Facial-expression-aware emotional color transfer	Pei et al. [7], Liu et al. [8]

and the source image. The PCA (Principal Component Analysis) technique is used to adjust the main trend or distribution axis of the two images in the RGB space and then match the distribution of the target image with that of the source image. Specifically, this method is performed as follows: Firstly, the covariance matrix of pixel values involved in the RGB space is summed. Then, the eigenvalues and eigenvectors of the matrix are determined. The eigenvector corresponding to the maximum eigenvalue represents the desired principal axis, which can be used to derive the rotation transformation of aligning the two color distributions. Finally, the adjusted RGB values are cut to a normal range.

3.1.2 Emotion Transfer Based on Emotion Word

Given an emotional word (e.g., enjoyable, pretty, heavy, etc.), Wang et al. [2] proposed an emotional word-based emotional color transfer method between images. They introduced color theme and devised the emotional word relationship model to select the best color theme from the candidate themes.

This emotion transfer algorithm is automatic and requires no user interaction. This method is composed of an offline stage and a running stage. In the offline stage, the relationship between the color theme and an emotional word is constructed by using the art theory and empirical concepts [10–13]. Emotional words and color themes are mapped to an image scale space. The 180 emotional words in the Kobayashi theory [12] have pre-defined image scale coordinates that are regarded as marker words. Here, the 5-color theme is employed because it can fully represent images with rich colors. By using artistic principles, the CC features (color composition theory guided feature descriptors) are designed, by which the color theme models with the image scale space are compared. Finally, Internet resources are exploited to expand the database to more than 400,000 color themes. In the running stage, given an emotional word, the standard semantic similarity is computed to obtain the weight corresponding to each marker word and get the coordinates in the image scale space. The closest candidate color theme is selected. Then, a strategy is designed to choose the most suitable color theme, balance the appearance compatibility with the target image, and adapt to the emotional word. Finally, the RBF (Radial Basis Function)-based interpolation method is utilized to adjust the image appearance according to the selected color theme.

3.1.3   Emotion Transfer Based on Color Combination

Color Combination Extraction. Color combination of an image, also known as color theme, refers to the color set that can reflect the main colors and color matching information in an image. The color theme extraction methods of images can be divided into three categories: clustering segmentation methods, statistical histogram methods, and data-driven methods.

The clustering segmentation method is a common color theme extraction method. An image is spatially clustered to obtain a locally clustered block image, so as to extract the colors representing each block, respectively. The K-means [14] and C-means [15] are the most used methods for clustering. The K-means method introduces the clustering number k, initially randomly selects K pixels as the clustering center, and iteratively transforms the clustering center to divide the image into k parts by minimizing the overall difference. The K-means method ignores the small color areas in an image, but one's favor for color is not ignored because of the proportion. The C-means method is similar to the K-means method, except that when minimizing the overall difference, it introduces a judgment on individual outliers to determine whether they belong to a new class or just individual outliers. The C-means method does not avoid the shortcomings of the K-means method in color theme extraction. Image segmentation methods [16] have also been applied to the extraction of color themes. Wang et al. [2] used the graph segmentation to divide an image into a small number of blocks. They scan and record the average pixel value of each block and then merge the blocks with similar pixel values, so as to achieve the required number of colors of a color theme.

The statistical histogram method stems from the distribution of pixel values in an image. Apart from spatial factors, only the number of occurrences of pixel values in an image is counted. The interval or point of pixel value aggregation in an image is obtained through statistical analysis as the color theme of an image. Delon et al. [17] converts an image into the HSV color space [18], counts the distribution histogram of H component, and divides the histogram into a region. The criterion of division is to divide according to the minimum value. Then, the region of H distribution is obtained, respectively, the S and V components are separately counted, so as to generate the main color of the given image. Morse et al. [19] controls a statistical amount when dividing, so as to reduce the number of colors. Zheng et al. [20] transformed the color space into three new emotional spaces and made histogram statistics according to the three emotional spaces. These methods count the color values that appear the most times. However, the theme one feels for an image cannot be simply determined by the number of occurrences.

The data-driven method is another commonly used color theme extraction method. Firstly, the algorithms mentioned in the above two categories are used to extract the color theme, and then the classifier is employed for optimization. Donovan et al. [21] extracts the color theme of an image by using the DERIECT algorithm [22]. On the other hand, according to the ranking of the themes on the image website, for the top color themes, the first-order linear LASSO regression model [23] is used to select important features. When optimizing the color theme, they select a series of similar color themes and rely on the ranking of the color theme in the database when determining the color theme. This method maintains the quality of color theme well, and the extracted color theme has high aesthetic value. However, this method pays too much attention to the original ranking of color themes and ignores one's preference for the connotation of an image. In addition, the color theme trained by this method suffers from over saturation, and the visual sense of color theme is less satisfactory.

Liu et al. [24] present an approach to extracting color themes from a given fabric image. They use a saliency map to judge the visual attention regions and separate the image into visual attention and non-visual attention regions. The dominant colors of the two regions are, respectively, computed and then merged to produce an initial target color theme. Without considering the emotion factors, the above color theme extraction methods can only produce a single color theme result for an image, which does not satisfy one's favor on different colors under different mood states. Liu and Luo [25, 26] presented a method to extract the emotional color theme from an image. They perform theme extraction with emotion value of each pixel rather than color value. The emotional discrepancy between colors in the theme is used to evaluate a color theme quality. Thus, this method establishes the emotional relationship between the target theme and the candidate theme.

Emotional Color Transfer Using Color Combination On the basis of the above color theme extraction methods, some researchers developed emotional color transfer approaches between images. He et al. [3] proposed an emotional color transfer framework based on color combination. The framework identifies three colors in an image by using the expectation-maximization clustering algorithm and models the color transfer process as two optimization problems, (1) calculating the target color combination and (2) maintaining the gradient of the target image. The user can select the target emotion by providing source images or directly selecting emotion keywords. If a source image is available, one can extract the main color (the clustering center in the color space) from the source image and use the closest scheme in the color emotion scheme as the target scheme. In this way, a specific scheme is selected and the color combination in the scheme is used for color transfer.

Meanwhile, the main color in the target image is also extracted. Once the main color in the target image is the same as the target scheme, all the colors in the target image are transferred to 24 output images (each scheme has 24 three-color combinations). Then, the best output image in the 24 three-color combinations is selected. Here, the aim of color transfer is to calculate the target color combination. The process of calculating the target color combination is formulated as an optimization problem:

$$\min_{\delta} f(\delta) = \sum_{k=1}^{3} w_k (\left\| C_I^k + \delta^k - C_{P_{ij}}^k \right\|_2^2)$$

$$s.t. \quad l\alpha\beta_{\min} \leq (I_{\min}^k + \delta^k) \leq l\alpha\beta_{\max},$$

$$l\alpha\beta_{\min} \leq (I_{\max}^k + \delta^k) \leq l\alpha\beta_{\max},$$

(3.1)

where $C_{P_{ij}}^k$ ($k = 1, 2, 3$) are three clustering centers of the target image calculated by the EM (Expectation-Maximum) algorithm. The δ^k represents the displacement of the cluster center, $l\alpha\beta_{min}$ and $l\alpha\beta_{max}$ define the range of each dimension in the $l\alpha\beta$ color space, I_{min}^k and I_{max}^k are the minimum and maximum values of the kth cluster in each dimension, and C_I^k ($k = 1, 2, 3$) is the number of colors for the target color combination.

Liu and Pei [4] presented a texture-aware emotional color transfer method between images, that can realize emotion transfer with an emotion word or a source image. Given an emotion word, this method can automatically change an image to the target emotion. Here, an emotion evaluation method is designed to compute the target emotion from a source image. The three-color emotion model databases are constructed to seek the proper color combinations expressing the target emotion. At last, a color transfer algorithm is designed to ensure the color gradient and naturalness by taking advantage of both color adjustment and color blending.

Figure 3.1 displays the pipeline of the above framework, which consists of three modules. Given a source image, this framework first extracts the main colors and texture features from the source image, which is used to calculate the target emotion coordinates with an emotion calculation model. If the input is an emotion word, a standard semantic similarity algorithm is utilized to estimate the matched word in the databases as the target emotion. Next, the most matching target emotion is searched in the emotion database and the closest color combination is selected from one of the model databases. They then obtain a target model with a special selection strategy and output the candidate color combinations. Finally, the target image is adjusted by color transfer.

3.2 Deep Learning-Based Emotional Color Transfer

Peng et al. [5] proposed a novel image emotion transfer framework based on CNN, in which they predict emotion distributions instead of simply predicting a single dominant emotion evoked by an image.

The emotional color transfer framework first shows that different people have different emotional changes to an image, i.e., one may also have a variety of different emotions for the same image. The framework uses the Emotion6 database to model emotion distribution. The Emotion6's emotion distribution predictor is used as a better baseline than previously used support vector regression (SVR) [2, 27, 28]. The framework also predicts emotions in the traditional environment of emotional image classification, indicating that CNN is superior

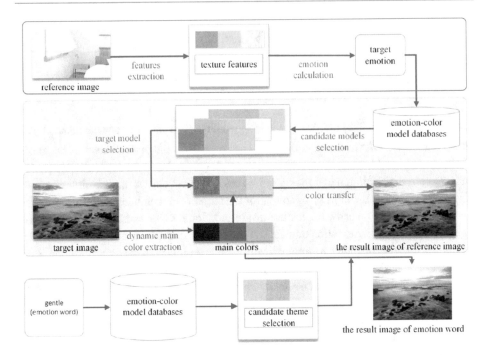

Fig. 3.1 The network structure of the texture-aware emotional color transfer method between images [4]

to Wang et al.'s method [2] in the artphoto dataset. Finally, with the support of a large-scale (600 image) user study, the framework successfully adjusts the emotional distribution of the image to the emotional distribution of the target image without changing the high-level semantics.

Liu et al. [6] proposed an emotional image color transfer method via deep learning. This method takes advantage of both the global emotion classification and local semantic for better emotional color enhancement. As shown in Fig. 3.2, this method is comprised of four ingredients, a shared low-level feature network (LLFN), an emotion classification network (ECN), a fusion network (FN), and a colorization network (CN). The LLFN takes in charge of extracting the semantic information. The ECN serves to constrain the color so that the enhancement results are more consistent with the source emotion. The ECN aims to treat the low-level features with four convolutional layers followed by fully connected layers. The ECN and the LLFN are then catenated by a fusion network. Finally, the CN is employed to create the emotional transferred results.

Facial-Expression-Aware Emotion Transfer. The existing deep learning methods do not consider the impact of face features on image emotion. When the background emotion in an image is inconsistent with the emotion expressed by the face in the image, the existing methods ignore the emotion expressed by the face, which will inevitably lead to the incon-

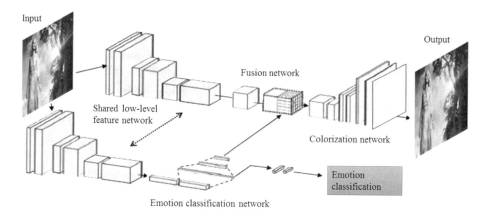

Fig. 3.2 The network structure of the emotional color enhancement framework based on deep learning [6]

sistency between the emotion of the resulting image and the emotion of the source image. For example, if the background of the target image is dark while the facial expression is happiness, previous methods would directly transfer dark color to the source image, neglecting the facial emotion in the image.

To this end, [7, 8] proposed an emotion transfer framework of face images based on CNN 3.3. The framework considers the face features of the image and accurately analyzes the face emotion of the image. Given a target face image, this method first predicts the facial emotion label of the image using an emotion classification network. The facial emotion labels are then matched with pre-trained emotional color transfer models. The pre-trained emotional models consist of 7 categories, namely the anger model, the disgust model, the happiness model, the fear model, the sadness model, the surprise model, and the neutral model. For example, if the predicted emotion label is "happy," then the corresponding emotion model is

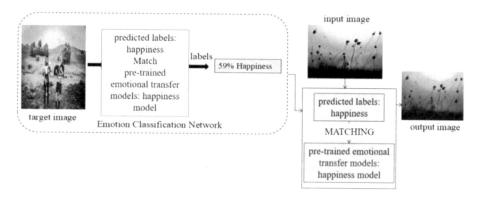

Fig. 3.3 The network structure of the facial-expression-aware emotional color transfer network [8]

the happiness model. The matched emotion model is used to transfer the color of the source image to the target image. Different from the previous methods, the emotion transfer model is obtained by a pre-training network. The network can simultaneously learn the global and local features of the image. Additionally, an image emotion database called "face emotion" is established. Different from the previous emotional image database, the images contained in the database are color images with facial expression features and obvious backgrounds, such as green mountains and rivers. The face emotion database provides useful data for facial emotion analysis.

References

1. C.K. Yang, L.K. Peng, Automatic mood-transferring between color images. IEEE Comput. Graph. Appl. **28**(2), 52–61 (2008)
2. X. Wang, J. Jia, L. Cai, Affective image adjustment with a single word. Vis. Comput. **29**(11), 1121–1133 (2013)
3. L. He, H. Qi, R. Zaretzki, Image color transfer to evoke different emotions based on color combinations. SIViP **9**(8), 1965–1973 (2015)
4. S. Liu, M. Pei, Texture-aware emotional color transfer between images. IEEE Access **6**, 31375–31386 (2018)
5. K.C. Peng, T. Chen, A. Sadovnik, A. Gallagheret, A mixed bag of emotions: model, predict, and transfer emotion distributions, in *Proceedings of IEEE Computer Vision and Pattern Recognition* (2015), pp. 860–868
6. D. Liu, Y. Jiang, M. Pei, S. Liu, Emotional image color transfer via deep learning. Pattern Recognit. Lett. **110**, 16–22 (2018)
7. M. Pei, S. Liu, and X. Zhang, Facial-expression-aware emotional color transfer based on convolutional neural network, in *Proceedings of the 26th Pacific Conference on Computer Graphics and Applications (PG) Posters* (2018), pp. 7–8
8. S. Liu, H. Wang, M. Pei, Facial-expression-aware emotional color transfer based on convolutional neural network. ACM Trans. Multimed. Comput. Commun. Appl. **18**(1), 8:1–8:19 (2022)
9. X. Xiao, L. Ma, Color transfer in correlated color space, in *Proceedings of ACM International Conference on Virtual Reality Continuum and Its Applications* (2006), pp. 305–309
10. R. Arnheim, *Art and Visual Perception: A Psychology of the Creative Eye*, University of California Press (1954)
11. J. Itten, *The Art of Color: the Subjective Experience and Objective Rationale of Color* (Wiley, New York, 1974)
12. S. Kobayashi, *Art of Color Combinations* (Kodansha International, Tokyo, 1995)
13. Y. Matsuda, *Color Design* (Asakura Shoten, Tokyo, 1995). ((in Japanese))
14. T.K. Means, S. Wang, E. Lien, Human toll-like receptors mediate cellular activation by mycobacterium tuberculosis. Comput. Graph. Forum **163**(7), 3920–3927 (1999)
15. R.J. Hathaway, J.C. Bezdek, Fuzzy C-means clustering of incomplete data. IEEE Trans. Syst. Man Cybern. B Cybern. **31**(5), 735–744 (2001)
16. H. Luo, S. Liu, Textile image segmentation through region action graph and novel region merging strategy, in *Proceedings of International Conference on Virtual Reality and Visualization (ICVRV)* (2014)

17. J. Delon, A. Desolneux, J.L. Lisani, Automatic color palette, in *Proceedings of IEEE International Conference on Image Processing* (2005), pp. 706–709
18. T.W. Chen, Y.L. Chen, S.Y. Chien, Fast image segmentation based on K-Means clustering with histograms in HSV color space, in *Proceedings of the 10th IEEE Workshop on Multimedia Signal Processing* (2008), pp. 322–325
19. B.S. Morse, D. Thornton, Q. Xia, Image-based color schemes, in *Proceedings of IEEE International Conference on Image Processing* (2007), pp. 497–500
20. D. Zheng, Y. Han, G. Baciu, Design through cognitive color theme: A new approach for fabric design, in *Proceedings of IEEE 11th International Conference on Cognitive Informatics & Cognitive Computing* (2012), pp. 346–355
21. O. Donovan, A. Agarwala, A. Hertzmann, Color compatibility from large datasets. ACM Trans. Graph. **30**(4), 63:1–63:12 (2011)
22. D.R. Jones, C.D. Perttunen, B.E. Stuckman, Lipschitzian optimization without the lipschitz constant. J. Optim. Theory Appl. **79**(1), 157–181 (1993)
23. R. Tibshirani, Regression shrinkage and selection via the lasso. J. Roy. Stat. Soc.: Ser. B (Methodol.) **58**(1), 267–288 (1996)
24. S. Liu, Y. Jiang, H. Luo, Attention-aware color theme extraction for fabric images. Text. Res. J. **88**(5), 552–565 (2018)
25. S. Liu, H. Luo, Hierarchical emotional color theme extraction. Color Res. Appl. **41**(5), 513–522 (2016)
26. S. Liu, H. Luo, Emotional color theme extraction, in *Proceedings of SIGGRAPH Asia Posters*, Article no. 27 (2014)
27. J. Machajdik, A. Hanbury, Affective image classification using features inspired by psychology and art theory, in *Proceedings of the International Conference on Multimedia* (2010), pp. 83–92
28. M. Solli, R. Lenz, Emotion related structures in large image databases, in *Proceedings of the ACM International Conference on Image and Video Retrieval* (2010), pp. 398–405

Colorization

There are a large number of grayscale or black-and-white images and video materials in various film, television, picture archives, medical, and other fields. Coloring them can greatly enhance the detail features and help one better identify and use them. Traditional manual coloring method consumes a lot of manpower and material resources and may not get satisfactory results. Given a source image or video, colorization methods aim to automatically colorize the target gray image or video reasonably and reliably, which thereby greatly improves the efficiency of this work.

4.1 Image Colorization

Colorization refers to adding colors to a grayscale image or video, which is a ill-posed task due to that it is ambiguous to assign colors to a grayscale image or video without any prior knowledge. So, at the early stage, user intervention is usually involved in image colorization. Later, automatic image colorization methods and deep learning learning-based colorization methods emerged.

4.1.1 Semi-automatic Colorization

Semi-automatic colorization methods require some amount of user interactions. Among them, color transfer methods [1–3] and image analogy methods [4] in Chap. 2 are widely used. In this case, a source image is provided as an example for coloring a given grayscale image, i.e., the target image. When the source image and the grayscale image share similar contents, impressive colorization results can be achieved. Nevertheless, these methods are labor intensive, since the source image and the target image should be manually matched.

© The Author(s), under exclusive license to Springer Nature Switzerland AG 2023
S. Liu, *Image and Video Color Editing*, Synthesis Lectures on Visual Computing: Computer Graphics, Animation, Computational Photography and Imaging,
https://doi.org/10.1007/978-3-031-26030-8_4

A luminance keying-based method for transferring color to a grayscale image is described in Gonzalez and Woods [5]. Color and grayscale values are matched with a pre-defined look-up table. When assigning different colors for the same gray level, a few luminance keys should be simultaneously manipulated by the user for different regions, making it a tiresome process. As an extension of color transferring method between color images [2], Welsh et al. [1] proposed transferring from a source color image to a target grayscale image. It matches color information between the two images with swatches.

Levin et al. [6] presented an efficient colorization method which allows users to interact with a few scribbles. With the observation that neighboring pixels in space-time share similar intensities and should have similar colors, they formulate colorization as an optimization problem for a quadratic cost function. With a few color scribbles by the user, the indicated colors can be automatically propagated in the grayscale image. Nie et al. [7] developed a colorization method by a local correlation-based optimization algorithm. This method depends on the color correlativity between pixels in different regions, limiting its practical application. Nie et al. [8] presented an efficient grayscale image colorization method. This method achieved comparable colorization quality with [6] with much less computation time by quadtree decomposition-based non-uniform sampling. Furthermore, this method greatly reduces the problem of color diffusion among different regions via designing a weighting function to represent intensity similarity in the cost function. This is an interactive colorization method, where the user needs to provide color hints by scribbling or seed pixels.

Irony et al. [9] presented a novel colorization method by transferring color from an exampler image. This method uses a strategy of the higher-level context of each pixel instead of independent pixel-level decisions in order to achieve better spatial consistency in the colorization result. Specifically, with a supervised classification scheme, they estimate the best example segment for each pixel to learn color from. Then, by combining a neighborhood matching metric and a spatial filter for high spatial coherence, each pixel is assigned a color from the corresponding region in the example image. It is reported that this approach requires considerably less scribbles than previous interactive colorization methods (e.g., [6]).

Yatziv and Sapiro [10] proposed an image colorization method via chrominance blending. This scheme is based on the concept of color blending derived from a weighted distance function that is computed from the luminance channel. This method is fast and allows the user to interactively colorize a grayscale image by providing a reduced set of chrominance scribbles. This method can also be extended for recolorization and brightness change.

Image color can be viewed as a highly correlated vector space. Abadpour and Kasaei [3] realized grayscale colorization by applying the PCA (Principal Component Analysis)-based transformation. They propose a category of colorizing methods that generate the color vector corresponding to the grayscale as a function. This method is significantly faster than previous approaches while producing visually acceptable colorization results. It can also be extended for recolorization. Nevertheless, this method is restricted by complicated segmentation that is tiresome by using the *magic select* tool in *Adobe Photoshop*. Luan et al. [11] proposed an interactive system for users to easily colorize natural images. The colorization procedure consists of two stages: color labeling and color mapping. In the first stage, pixels that should have similar colors are grouped into coherent regions in the first stage, while in the second stage, color mapping is applied to generate a vivid colorization effect by assigning colors to a few pixels in the region. It is very tedious to colorize texture by previous methods since each tiny region inside the texture needs a new color. In contrast, this method handles texture by grouping both neighboring pixels sharing similar intensity and remote pixels with a similar texture. This method is effective for natural image colorization. However, the user should usually provide multiple stokes on similar patterns with different orientations and scales in order to produce fine colorization results.

Liu et al. [12] proposed an example-based colorization method that is aware of illumination differences between the target grayscale image and the source color image. Firstly, an illumination-independent intrinsic reflectance map of the target scene is recovered from multiple color references collected by web search. Then, the grayscale versions of the reference images are employed for decomposing the target grayscale image into its intrinsic reflectance and illumination components. The color is transferred from the color reflectance map to the grayscale reflectance image. By relighting with the illumination component of the target image, the final colorization result can be produced. This method needs to search suitable source images for reference by web search.

Liu et al. [13] presented a grayscale image colorization method by control of a single parameter. The polynomial fitting models of the histograms of the source image and the grayscale image are computed by linear regression, respectively. With the user-assigned order of the polynomials, the source image and the grayscale images can be automatically matched. By transferring between the corresponding regions of the source image and the grayscale image, colorization can be finally achieved. Quang et al. [14] proposed an image and video colorization method based on the kernel Hilbert space (RKHS). This method can produce impressive colorization results. Nevertheless, it requires initialization for different regions by manual, which is time-consuming if there are many different contents in the grayscale image.

4.1.2   Automatic Colorization

The above colorization methods require the user to perform colorization by manual, either providing a source image or using scribbles and color seeds for interaction. Since there is usually no suitable correspondence between color and local texture, automatic colorization is necessary.

Li and Hao [15] proposed an automatic colorization approach by locally linear embedding. Given a source color image and a target grayscale image, this method clips both of them into overlapping patches, which are supposed to be distributed on a manifold [16, 17]. For each patch, its neighborhood in the training patches is estimated and its chromatic information is predicted by the manifold learning [18]. With multi-modality, Charpiat et al.

[19] predict the probability distribution of all possible colors for each pixel of the image to be colored, rather than selecting the probable color locally. Then, the technique of graph cut is employed to maximize the probability of the whole colored image globally. Morimoto et al. [20] proposed an automatic colorization method using multiple images collected from the web. Firstly, this method chooses images with similar scene structure with the target grayscale image I_m as the source images. The *gist scene descriptor*, a feature vector expressing the global scene in a lower dimension is used to aggregate oriented edge responses at multi-scales into coarse spatial bins. Then the distance between the gist of I_m and that of the images from the web is computed. The most similar images are chosen as source images, which are used for colorization. Here, the transferring method of [1] was used for colorization. However, this method restricts the searching results from the images collected from the web, which may produce unnatural results due to the source images that are structurally similar but semantically different.

To this end, Liu and Zhang [21] proposed an automatic grayscale image colorization method via histogram regression. Given a source image and a target grayscale image, the locally weighted regression is performed on both images to obtain the feature distributions of them. Then, these features are automatically matched by aligning zero points of the histogram. Thus, the grayscale image is colorized in a weighted manner. Figure 4.1 shows a colorization result by this method. Although this method can achieve confident colorization results, it may fail for images with strong texture patterns or varied lighting effects (e.g., shadows and highlights).

Liu and Zhang [22] further proposed a colorization method based on a texture map. Assuming that a source color image with the similar content to the target grayscale image can be provided by the user, this method is aware of both the luminance and texture information of images so that more convincing colorization results can be produced. Specifically, given a source color image and a target grayscale image, their respective spatial maps are computed. Note that the spatial map is a function of the original image, indicating the luminance spatial distribution for each pixel. Then, by performing locally weighted linear regression on the histogram of the quantized spatial map, a series of spatial segments are computed. Within each segment, the luminance of the target grayscale image is automatically mapped to color values. Finally, colorization results can be yielded through local luminance-color correspondence and global luminance-color correspondence between the source color image and the target grayscale image.

Beyond natural images, Visvanathan et al. [23] automatically colorized pseudocolor images by gradient-based value mapping. This method targets visualizing pixel values and their local differences for scientific analysis.

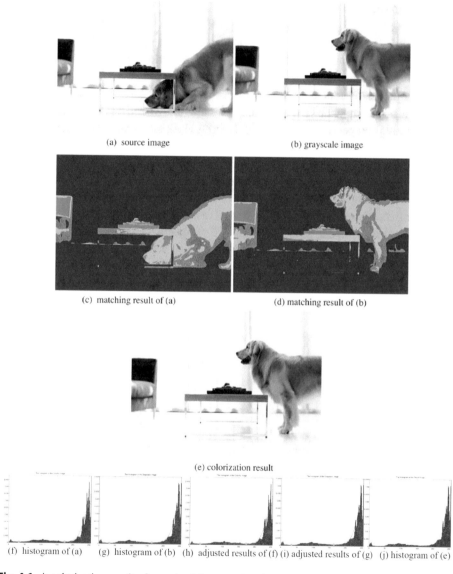

(a) source image   (b) grayscale image

(c) matching result of (a)   (d) matching result of (b)

(e) colorization result

(f) histogram of (a)   (g) histogram of (b)   (h) adjusted results of (f)   (i) adjusted results of (g)   (j) histogram of (e)

Fig. 4.1 A colorization result of an animal image using the automatic grayscale image colorization method via histogram regression [21]

4.2   Cartoon Colorization

Some researchers also extended the colorization technique to cartoon images. Sýkora et al. [24] proposed a semi-automatic, fast, and accurate segmentation method for black-and-white cartoons. It allows the user to efficiently apply ink on the aged black-and-white cartoons. The inking process is composed of four stages, namely segmentation, marker prediction, color luminance modulation, and final composition of foreground and background layers.

Qu et al. [25] proposed a method for colorizing black-and-white manga (comic books in Japanese) containing a large number of strokes, hatching, halftoning, and screening. Given scribbles by the user on the target grayscale manga drawing, Gabor wavelet filters are employed to measure the pattern continuity, and thereby a local, statistical-based pattern feature can be estimated. Then, with the level set technique, the boundary is propagated to monitor the pattern continuity. In this way, areas with open boundaries or multiple disjointed regions with similar patterns can be well segmented. Once the segmented regions are obtained, conventional colorization methods can be used to color replacement, color preservation as well as pattern shading.

Table 4.1 gives a comparison between different colorization methods in terms of the color source, degree of automation, interaction style, texture-aware or not, and their application domain.

Table 4.1 A comparison between different colorization methods

Methods	Color Source	Automatic	Interactions	Texture-aware	Application
[1]	Source image	No	Swatches	No	Natural image
[6]	Source image	No	Strokes	No	Natural image
[9]	Example image	No	Scribbles	No	Natural image
[10]	Source image	No	Scribbles	No	Natural image
[11]	User specified color	No	Strokes	Yes	Natural image
[25]	User specified color	No	Scribbles	Yes	Manga or cartoons
[26]	User specified color	No	Scribbles	Yes	Natural image
[27]	Source color imagery	No	Swatches	No	Aerial/space imagery
[21]	Source image	Yes	No	No	Natural image
[23]	Pseudo-color	Yes	No	No	Data image
[19]	Source image	Yes	No	yes	Natural image
[20]	Web images	Yes	No	No	Natural image
[22]	Source image	Yes	No	Yes	Natural image

4.3 Deep Colorization

Cheng et al. [28] proposed a deep neural network model to achieve fully automatic image colorization by leveraging a large set of source images from different categories (e.g., animal, outdoor, indoor) with various objects (e.g., tree, person, panda, and car). This method consists of two stages, (1) training a neural network and (2) colorizing a target grayscale image with the learned neural network.

Larsson et al. [29] trained a model to predict per-pixel color histograms for colorization. This method trains a neural architecture in an end-to-end manner by considering semantically meaningful features of varying complexity. Then, a color histogram prediction framework is developed to treat uncertainty and ambiguities inherent in colorization so as to avoid jarring artifacts. Given a grayscale image, with a deep convolutional architecture (VGG), spatially localized multilayer slices are chosen as per-pixel descriptors. The system then estimates hue and chroma distributions for each pixel p with its hypercolumn descriptor. Finally, at test time, the estimated distributions are used for color assignment.

Zhang et al. [30] treat image colorization as a classification problem considering the underlying uncertainty of this task. They leverage class-rebalancing during training to increase the diversity of colors. At test time, this method is performed as a feed-forward pass in a CNN with million color images. This method demonstrates that with a deep CNN and a carefully-tuned loss function, the colorization task can generate results close to real color photos.

Iizuka et al. [31] proposed an automatic, CNN-based grayscale image colorization method by combining both global priors and local image features. The proposed network architecture is able to jointly extract global and local features from an image and fuse them for colorization. Specifically, their model is composed of four parts, namely a low-level features network, a mid-level features network, a global features network, and a colorization network. Various evaluation experiments were performed to verify this method with a user study and many historical hundred-year-old black-and-white photographs.

Zhang et al. [32] propose a CNN framework for user-assisted image colorization. Given a target grayscale image, and sparse, local user edits, this method can automatically produce convincing colorization results. By training on a large amount of image data, this method learns to propagate user edits by merging both low-level cues and high-level semantic information. This method has helped non-professionals to design a colorful work, since it has a great ability to achieve fine colorization results even with random user inputs.

Deshpande et al. [33] learned a low-dimensional smooth embedding of color fields with a variational autoencoder (VAE) for grayscale image colorization. A multi-modal conditional model between the gray-level features and the low-dimensional embedding is learned to produce *diverse* colorization results. The loss functions are specially designed for the VAE decoder to avoid blurry colorization results and respect the uneven distribution of pixel colors. This method has the potential to handle other ambiguous problems, since the low-dimensional embeddings have the ability to predict diversity with multi-modal conditional models. However, high spatial detail is not taken into account in this method.

4.4   Video Colorization

People are willing to watch a colorful film instead of a grayscale one. *Gone with the Wind* in 1939 is one of the first colorized films [34] which is popular with the audience. However, it is challenging to obtain a convincing video colorization because of its multi-modality in the solution space and the requirement of global spatiotemporal consistency [35] is also inherently more challenging than Unlike single-image colorization, video colorization should also satisfy temporal coherence. In view of this point, the above single-image colorization cannot be used for video colorization. Currently, researchers [35–39] realized video colorization by propagating the color information either from a color reference frame or sparse user scribbles to the whole target grayscale video.

Vondrick et al. [37] regard video colorization as a self-supervised learning problem for visual tracking. To this end, they learn to colorize grayscale videos by copying colors from a reference frame by exploiting the temporal coherency of color, rather than predicting the color directly from the grayscale frame. Jampani et al. [36] proposed Video Propagation Network (VPN) that processes video frames in an adaptive manner. The VPN consists of a temporal bilateral network (TBN) and a spatial network (SN). The TBN aims for dense and video adaptive filtering, while the SN is used for refining features and increasing flexibility. This method propagates information forward without accessing future frames. Experiments showed that, given the source color image for the first video frame, this method can propagate the color to the whole target grayscale video. Given the color image for the first video frame, the task of this method is to propagate the color to the entire video. This method can also be used for video processing tasks requiring the propagation of structured information (e.g., video object segmentation and semantic video segmentation).

Meyer et al. [39] proposed a deep learning framework for video color propagation. This method consists of a short-range propagation network (SRPN), a longer-range propagation network (LRPN), and a fusion and refinement network (FRN). The SRPN aims to propagate colors frame by frame ensuring temporal stability. The input to SRPN is two consecutive grayscale frames and it outputs an estimated warping function that is used to transfer the colors of the previous frame to the next one. The LRPN introduces semantical information by matching deep features extracted from the frames, which are then used to sample colors from the first frame. Except for long-range color propagation, this strategy also contributes to restore missing colors because of occlusion. With a CNN, the SRPN is used to combine the above two stages for fusion and refinement.

Lei and Chen [35] proposed a fully automatic, self-regularized approach to video colorization with diversity. This method is comprised of a colorization network for video frame colorization and a refinement network for spatiotemporal color refinement. A diversity loss is designed to allow the network to generate colorful videos with diversity. Moreover, the diversity loss can also make the training and process more stable.

Table 4.2 summarizes the image and video colorization methods.

Table 4.2 Image and video colorization methods

Methods	References
Semi-automatic colorization	Gonzalez and Woods [5], Welsh et al. [1]
	Levin et al. [6], Nie et al. [7]
	Irony et al. [9], Yatziv and Sapiro [10]
	Abadpour and Kasaei [3], Nie et al. [8]
	Luan et al. [11], Liu et al. [12]
	Quang et al. [14], Liu et al. [13]
Automatic colorization	Visvanathan et al. 2007 [23], Li and Hao [15]
	Charpiat et al. [19], Morimoto et al. [20]
	Liu and Zhang [21], Liu and Zhang [22]
Cartoon colorization	Sýkora et al. [24], Qu et al. [25]
Deep colorization	Cheng et al. [28], Larsson et al. [29]
	Zhang et al. [30], Iizuka et al. [31]
	Deshpande et al. [33], Zhang et al. [32]
Video colorization	Jampani et al. [36], Meyer et al. [39]
	Vondrick et al. [37], Lei and Chen [35]

References

1. T. Welsh, M. Ashikhmin, K. Mueller, Transferring color to grayscale images, in *Proceedings of ACM SIGGRAPH*, pp. 277–280 (2002)
2. E. Reinhard, M. Ashikhmin, B. Gooch, P. Shirley, Color transfer between images. IEEE Comput. Graph. Appl. **21**(5), 34–41 (2001)
3. A. Abadpour, S. Kasaei, An efficient PCA-based color transfer method. J. Vis. Commun. Image Represent. **18**(1), 15–34 (2007)
4. A. Hertzmann, C.E. Jacobs, N. Oliver, et al., Image analogies, in *Proceedings of ACM SIG-GRAPH*, pp. 327–340 (2001)
5. B.C. Gonzalez, R.E. Woods, *Digital Image Processing*, 2nd edn. (Addison-Wesley Publishing, MS, 1987), p.1987
6. A. Levin, D. Lischinski, Y. Weiss, Colorization using optimization. ACM Trans. Graph. **23**(3), 689–694 (2004)
7. D. Nie, L. Ma, S. Xiao, X. Xiao, Grey-scale image colorization by local correlation based optimization algorithm, in *Proceedings of Visual*, pp. 13–23 (2005)
8. D. Nie, Q. Ma, L. Ma, S. Xiao, Optimization based grayscale image colorization. Pattern Recognit. Lett. **28**(12), 1445–1451 (2007)
9. R. Irony, D. Cohen-Or, D. Lischinski, Colorization by example, in *Proceedings of Eurographics Symposium on Rendering Techniques*, pp. 201–210 (2005)
10. L. Yatziv, G. Sapiro, Fast Image and video colorization using chrominance blending. IEEE Trans. Image Process. **15**(5), 1120–1129 (2006)

11. Q. Luan, F. Wen, D. Cohen-or, L. Liang, Y.-Q. Xu, H.-Y. Shum, Natural image colorization, in *Proceedings of the 18th Eurographics conference on Rendering Techniques*, pp. 309–320 (2007)
12. X. Liu, L. Wan, Y. Qu, T. Wong, L.S. Lin, P.A. Heng, Intrinsic colorization. ACM Trans. Graph. **27**(5), 152:1–152:9 (2008)
13. S. Liu, X. Zhang, J. Wu, J. Sun, Q. Peng, Gray-scale image colorization based on the control of sing-parameter. J. Image Graph. **16**(7), 1297–1302 (2011)
14. M. Quang, H. Kang, T.M. Le, Image and video colorization using vector-valued reproducing kernel Hilbert spaces. J. Math. Imaging Vis. **37**(1), 49–65 (2010)
15. J. Li, P. Hao, Transferring colours to grayscale images by locally linear embedding, in *Proceedings of British Machine Vision Conference (BMVC)*, pp. 835–844 (2008)
16. W. Fan, D.-Y. Yeung, Image hallucination using neighbor embedding over visual primitive manifolds, in *Proceedings of IEEE Conference on Computer Vision and Pattern Recognition (CVPR)* (2007)
17. D. Beymer, T. Poggio, Image representation for visual learning. Science **272**(5270), 1905–1909 (1996)
18. S.T. Roweis, L. Saul, Nonlinear dimensionality reduction by locally linear embedding. Science **290**(5500), 2323–2326 (2000)
19. G. Charpiat, M. Hofmann, B. Scholkopf, Image colorization via multimodal predictions, in *Proceedings of the 10th European Conference on Computer Vision (ECCV)*, pp. 126–139 (2008)
20. Y. Morimoto, Y. Taguchi, T. Naemura, Automatic colorization of grayscale images using multiple images on the web, in *Proceedings of ACM SIGGRAPH Posters*, Article no. 32 (2009)
21. S. Liu, X. Zhang, Automatic grayscale image colorization using histogram regression. Pattern Recognit. Lett. **33**(13), 1673–1681 (2012)
22. S. Liu, X. Zhang, Image colorization based on texture map. J. Electron. Imaging **22**(1), 013011:1–9 (2013)
23. A. Visvanathan, S.E. Reichenbach, Q. Tao, Gradient-based value mapping for pseudocolor images. J. Electron. Imaging **16**(3), Article no. 033004 (2007)
24. D. Sýkora, J. Buriánek, J. Žára, Segmentation of black and white cartoons, in *Proceedings of Spring Conference on Computer Graphics*, pp. 245–254 (2003)
25. Y. Qu, T. Wong, P.A. Heng, Manga colorization. ACM Trans. Graph. **25**(3), 1214–1220 (2006)
26. M. Kawulok, B. Smolka, Texture-adaptive image colorization framework. EURASIP J. Adv. Signal Process (1), 99:1–99:15 (2011)
27. U. Lipowezky, Grayscale aerial and space image colorization using texture classification. Pattern Recognit. Lett. **27**(4), 275–286 (2006)
28. Z. Cheng, Q. Yang, B. Sheng, Deep colorization, in *Proceedings of the IEEE International Conference on Computer Vision (ICCV)*, pp. 415–423 (2015)
29. G. Larsson, M. Maire, G. Shakhnarovich, Learning representations for automatic colorization, in *Proceedings of the European Conference on Computer Vision (ECCV)*, pp. 577–593 (2016)
30. R. Zhang, P. Isola, A.A. Efros, Colorful image colorization, in *Proceedings of European Conference on Computer Vision (ECCV)*, pp. 649–666 (2016)
31. S. Iizuka, E. Simo-Serra, H. Ishikawa, Let there be Color!: joint end-to-end learning of global and local image priors for automatic image colorization with simultaneous Classification. ACM Trans. Graph. **35**(4), 110:1–110:11 (2016)
32. R. Zhang, J. Zhu, P. Isola, X. Geng, A.S. Lin, T. Yu, A.A. Efros, Real-time user-guided image colorization with learned deep priors. ACM Trans. Graph. **36**(4), 119:1–119:11 (2017)
33. A. Deshpande, J. Lu, M. Yeh, M.J. Chong, D.A. Forsyth, Learning diverse image colorization, in *Proceedings of IEEE Conference on Computer Vision and Pattern Recognition (CVPR)*, pp. 2877–2885 (2017)

34. Highest-grossing film at the global box office (inflation-adjusted)—guinness world records. http://www.guinnessworldrecords.com/worldrecords/highest-box-office-film-gross-inflation-adjusted
35. C. Lei, Q. Chen, Fully automatic video colorization with self-regularization and diversity, in *Proceedings of IEEE/CVF Conference on Computer Vision and Pattern Recognition (CVPR)*, pp. 3753–3761 (2019)
36. V. Jampani, R. Gadde, P.V. Gehler, Video propagation networks, in *Proceedings of IEEE Conference on Computer Vision and Pattern Recognition (CVPR)* (2017)
37. C. Vondrick, A. Shrivastava, A. Fathi, S. Guadarrama, K. Murphy, Tracking emerges by colorizing videos, in *Proceedings of European Conference on Computer Vision (ECCV)*, pp. 402–419 (2018)
38. S. Liu, G. Zhong, S.D. Mello, J. Gu, M. Yang, J. Kautz, Switchable temporal propagation network, in *Proceedings of European Conference on Computer Vision (ECCV)* (2018)
39. S. Meyer, V. Cornillère, A. Djelouah, C. Schroers, M.H. Gross, Deep video color propagation, in *Proceedings of the International Conference on Multimedia*, pp. 83–92 (2018)

Decolorization

Image or video decolorization, also known as grayscale transformation, converts a three-channel color image or video into a single-channel grayscale one. Decolorization is actually a process of *dimension reduction*, so that the resulting grayscale image or video often only contains the most important information, which greatly saves storage space. A grayscale image or video can better display the texture and contour of objects. Decolorization can also be widely applied in the field of image compression, medical image visualization, and image or video art stylization. Black-and-white digital printing of images, with the advantages of low cost and fast printing, is common in daily life, one important process of which is decolorization, i.e., a color image sent to a monochrome printer must undergo a color-to-grayscale transformation. Below we will summarize various image and video decolorization methods in the last decade.

5.1   Image Decolorization

Image decolorization is often used as a preprocessing for downstream image processing tasks such as segmentation, recognition, and analysis. Recently, decolorization has attracted more and more attention from researchers. In the early stage, the three channels R, G, and B are represented by a single channel or only the brightness channel information is used to represent the grayscale image. However, these simple color removal methods suffer from contrast loss in the gray image. To this end, researchers have proposed local and global decolorization methods in order to preserve the contrast of color images in the resulting grayscale images.

S. Liu, *Image and Video Color Editing*, Synthesis Lectures on Visual Computing: Computer Graphics, Animation, Computational Photography and Imaging,
https://doi.org/10.1007/978-3-031-26030-8_5

43

5.1.1   Early Decolorization Methods

The early image decolorization method is simple, which directly processes the (R, G, B) channels of a color image in the RGB color space. These methods include the component method, the maximum method, the average method, and the weighted average method.

The component method uses one of the (R, G, B) in the color image as the corresponding pixel value in the grayscale image, written as

$$G_1(i, j) = R(i, j),$$
$$G_2(i, j) = G(i, j), \tag{5.1}$$
$$G_3(i, j) = B(i, j),$$

where (i, j) is the pixel coordinate in an image. Note that any one of G_1, G_2, G_3 can be selected as needed.

The maximum method takes the maximum value of (R, G, B) in the color image as the gray value of the grayscale image.

$$GRAY(i,j) = \max\{R(i, j), G(i, j), B(i, j)\}. \tag{5.2}$$

The average method is to average the three component values of (R, G, B) in the color image to obtain a gray value.

$$GRAY(i,j) = (R(i, j), G(i, j), B(i, j))/3. \tag{5.3}$$

The weighted average method uses the weighted average of three components with different weights as the grayscale image.

$$GRAY(i,j) = 0.299R(i, j) + 0.578G(i, j) + 0.114B(i, j). \tag{5.4}$$

In addition to using the color component of the RGB space, it is also common to employ the brightness channel of other color spaces to represent the gray value of a grayscale image. For example, Hunter [1] uses the L channel of the $L\alpha\beta$ space to represent a grayscale image, while Wyszecki and Stiles [2] adopt the Y component in the YUV color space to represent the grayscale image. In the YUV color space, the Y component is the brightness of pixels, reflecting the brightness level of an image. According to the relationship between the RGB color space and the YUV color space, the mapping between the brightness y and three color components can be established as

$$y = 0.3r + 0.59g + 0.11b. \tag{5.5}$$

The luminance value y is used to represent the gray value of the image. Based on this observation, Nayatani [3] proposed a color mapping model with independent input, i.e., input three components independently and set the weights of the corresponding components as needed.

These early methods are easy to implement; however, they would cause the loss of image contrast, saturation, exposure, etc. To this end, researchers explored decolorization methods with higher accuracy and efficiency, including local decolorization methods, global decolorization methods, and deep learning-based decolorization methods.

5.1.2 Local Decolorization Methods

Local decolorization methods usually use different strategies in solving the mapping model from a color image to a grayscale one. The strategy deals with different pixels or color blocks and increases the local contrast by strengthening the local features.

Bala and Eschbach [4] proposed a decolorization method that locally enhances the edge and contours between adjacent colors by adding high-frequency chrominance information into the luminance channel. Specifically, a spatial high-pass filter weighting the output with a luminance-dependent term is applied to the chrominance channels. Then the result is added to the luminance channel.

Neumann et al. [5] view the color and luminance contrasts of an image as a gradient field and solve the inconsistency of the field. They chose locally consistent color gradients and performed 2D integration to produce the grayscale image. Since its complexity is linear in the number of pixels, this method is simple yet very efficient, which is suitable for handling high-resolution images. Smith et al. [6] proposed a perceptually accurate decolorization method for both images and videos. This approach consists of two steps: (1) globally assigning gray values and determining color ordering and (2) locally improving the grayscale to preserve the contrast in the input color image. The Helmholtz-Kohlrausch color appearance effect is introduced to estimate distinctions between iso-luminant colors. They also designed a multi-scale local contrast enhancement strategy to produce a faithful grayscale result. Note that this method makes a good balance between a fully automatic method (first step) and user assist (second step), making it suitable for dealing with various images (e.g., natural images, photographs, artistic works, and business graphics). For a challenging image consisting of equiluminant colors, this method is able to predict the H-K effect that makes a more colorful blue appear lighter than the duller yellow. A limitation of this approach comes from the locality of the second step, which may fail to preserve chromatic contrast between non-adjacent regions and lead to temporal inconsistencies.

Lu et al. [7] proposed a decolorization method aiming to preserve the original color contrast as far as possible. A bimodal contrast preserving function is designed to constrain local pixel differences and a parametric optimization approach is employed to preserve the original contrast. Owing to weak color order constraints, they relax the color order constraint and seek to better maintain color contrast and enhance the visual distinctiveness of edges. Nevertheless, this method cannot greatly preserve the global contrast in the image. Moreover, since the gray image is produced by solving the energy equation in an iterative manner, the efficiency of this algorithm is relatively low. Zhang and Liu [8] presented an efficient image decolorization method via perceptual group difference (PGD) enhancement. They view the perceptual group instead of individual image pixels as the human perception elements. Based on this observation, they perform decolorization for different groups in order to maximumly maintain the contrast between different visual groups. A global color-to-gray mapping is employed to estimate the grayscale of the whole image. Experimental results showed that, with PGD enhancement, this approach is capable of achieving better visual contrast effects.

The local decolorization methods may distort the appearance of regions with constant colors and therefore lead to undesired haloing artifacts.

5.1.3   Global Decolorization Methods

Global decolorization methods perform decolorization on the whole image in a global manner, including linear decolorization and nonlinear decolorization techniques.

Linear decolorization methods. Gooch et al. [9] proposed Color2Gray, a saliency-preserving decolorization method. This method is performed in the CIE $L^*a^*b^*$ color space instead of the traditional RGB color space. Considering that the human visual system is sensitive to change, they preserve relationships between neighboring pixels rather than representing absolute pixel values. The chrominance and luminance changes in a source image are transferred to changes in the target grayscale image so as to produce images maintaining the salience of the source color images. Grundland and Dodgson [10] proposed an efficient, linear decolorization approach by adding a fixed amount of chrominance to lightness. To achieve a perceptually plausible decolorization result, Kuk et al. [11] proposed a color-to-grayscale conversion method by taking into account both local and global contrast. They encode both local and global contrast into an energy function via a target gradient field, which is constructed from two types of edges: (1) edges connecting each pixel to neighboring pixels, and (2) edges connecting each pixel to predetermined landmark pixels. Finally, they formulate the decolorization problem as reconstructing a grayscale image from the gradient field, which is solved by a fast 2D Poisson solver.

Nonlinear declorization methods. Kim et al. [12] presented a fast and robust decolorization algorithm via a global mapping that is a *nonlinear* function of the lightness, chroma, and hue of colors. Given a color image, the parameters of the function are optimized to make the resulting grayscale image respect the feature discriminability, lightness, and color ordering in the input color image. Ancuti et al. [13] introduced a fusion-based decolorization technique. The input of their method includes three independent RGB channels and an additional image that conserves the color contrast. The weights are based on three different forms of local contrast: a saliency map to preserve the saliency of the original color image, a second weight map taking advantage of well-exposed regions, and a chromatic weight map enhancing the color contrast. By enforcing a more consistent gray-shade ordering, this strategy can better preserve the global appearance of the image. Ancuti et al. [14] further presented a color-to-gray conversion method aiming to enhance the contrast of the images while preserving the appearance and quality of the original color image. They intensify the monochromatic luminance with a mixture of saturation and hue channels in order to respect the original saliency while enhancing the chromatic contrast. In this way, a novel spatial distribution can be produced which is capable of better discriminating the illuminated regions and color features. Liu et al. [15] developed a decolorization model based on gradient correlation similarity (GCS) so as to reliably maintain the appearance of the source color image. The gradient correlation is employed as a criterion to design a *nonlinear* global mapping in the RGB color space. This method is able to better preserve features in the source color image which are more discriminable in the grayscale image, and it also has a good ability to maintain a desired color ordering in color-to-gray conversion. Liu et al. [16] further proposed a color-to-grayscale method by introducing the gradient magnitude [17].

Song et al. [18] regard decolorization as a labeling problem to maintain the visual cues of a color image in the resulting grayscale image. They define three types of visual cues, namely color spatial consistency, image structure information, and color channel perception priority, that can be extracted from a color image. Then, they cast color to gray as a visual cue preservation process based on a probabilistic graphical model, which is solved via integral minimization.

Most of the above image decolorization methods attempt to preserve as much as possible visual appearance and color contrast; however, little attention was devoted to the speed issue of decolorization. The efficiency of most methods is lower than the standard procedure (e.g., Matlab built-in *rgb2gray* function). To this end, Lu et al. [19] proposed a *real-time* contrast preserving decolorization method. They achieved this goal with three main ingredients: a simplified bimodal objective function with a linear parametric grayscale model, a fast non-iterative discrete optimization, and a sampling-based P-shrinking optimization strategy. The running time of this method is a constant $O(1)$, independent of image resolutions. This method takes only 30ms to decolorize an one megapixel color image, which is comparable with the built-in Matlab *rgb2gray* function, but achieves a better color-to-gray conversion result which is visually similar to a compelling contrast preserving decolorization method [7].

Lu et al. [20] further presented an optimization framework for image decolorization to preserve color contrast in the original color image as much as possible. A bimodal objective function is used to reduce the restrictive order constraint for color mapping. Then, they design a solver to realize the automatic selection of suitable grayscales via global contrast constraints. They also propose a quantitative perceptual-based metric, E-score, to measure contrast loss and content preservation in the resulting grayscale images. The E-score is to jointly consider two measures CCPR (Color Contrast Preserving Ratio) and CCFR (Color Content Fidelity Ratio), written as

$$E_{score} = \frac{2 \cdot CCPR \cdot CCFR}{CCPR + CCFR}. \tag{5.6}$$

It is reported that this is among the first attempts in the color-to-gray field to quantitatively evaluate decolorization results.

Considering that the above decolorization methods suffer from the robustness problem, i.e., may fail to accurately convert iso-luminant regions in the original color image, while the $rgb2gray()$ function in Matlab works well in practice applications. Song et al. [21] proposed a robust decolorization method by modifying the $rgb2gray()$ function. This method is able to realize color-to-gray conversion for iso-luminance regions in an image, while previous methods, including Gooch et al. [9], Gundland and Dodgson [10], Smith et al. [6], Kim et al. [12], Ancuti et al. [14], and Lu et al. [7, 19] fail in this task. In this method, they avoid indiscrimination in iso-luminant regions by adaptively selecting channel weights with respect to specific images rather than using fixed channel weights for all cases. Therefore, this method is able to maintain multi-scale contrast in both spatial and range domains.

Sowmya et al. [22] presented a color-to-gray conversion algorithm with a weight matrix corresponding to the chrominance components. The weight matrix is obtained by reconstructing the chrominance data matrix through singular value decomposition (SVD). Ji et al. [23] presented a global image decolorization approach with a variant of the difference-of-Gaussian band-pass filter, called luminance filters. Typically, the filter has high responses on regions in which colors differ from their surroundings for a certain band. Then, the grayscale value can be produced after luminance passing a series of band-pass filters. Due to that this approach is linear in the number of pixels, it is efficient and easy to implement.

5.1.4 Deep Learning-Based Decolorization Methods

The goal of the deep learning-based image decolorization methods is to train a nonlinear mapping model from a color image to a gray image. There are usually two main steps: (1) training the neural network with a large number of input-output images to obtain the mapping model from the trained color image to the gray image; (2) The learned neural network model is used to decolorize the target color image. The data input to the convolution neural network is a three-channel color image c, and the output is a single-channel gray image g,

$$g = \sigma(c) \tag{5.7}$$

where σ represents the nonlinear mapping function to be learned.

By training partial differential equations (PDEs) on 50 input/output image pairs, Lin et al. [24] constructed a mapping model for the task of color-to-gray conversion. It is reported that their learned PDEs can yield similar decolorization results to those of Gooch et al. [9].

Hou et al. [25] proposed the Deep Feature Consistent Deep Image Transformation (DFC-DIT) framework for one-to-many mapping image processing tasks (e.g., downscaling, decolorization, and tone mapping). The DFC-DIT achieves transformation between images with a CNN as a nonlinear mapper respecting the deep feature consistency principle that is enforced with another pre-trained and fixed deep CNN. This system is comprised of two networks, a transformation network and a loss network. The former is used to convert an input to an output, and the latter serves as computing the feature perceptual loss for the training of the transformation network.

Considering that the local decolorization methods are less accurate enough to process local pixels leading to local artifacts, while the global methods may fail to treat local color blocks, Zhang and Liu [26] proposed a novel image color-to-gray conversion method by combining local semantic features and global features. In order to preserve color contrast between adjacent pixels, a global feature network is developed to learn the global features and spatial correlation of an image. On the other hand, in order to preserve the contrast between different object blocks, they take care of local semantic features of images and fine classification of pixels during learning deep image features. Finally, with the fusion of both the local semantic features and global features, this method performs better in terms of contrast preservation than the state-of-the-art decolorization approaches. Figure 5.1 gives a flow chart of this method.

According to the human visual mechanism, exposure plays a critical role in human visual perception, e.g., low-exposure and over-exposure areas usually easily catch the attention of an observer. However, exposure is missed in existing decolorization methods. To this end, Liu and Zhang [27] proposed an image decolorization approach by fusion of local features and exposure features with a CNN framework. This framework consists of a local feature network and a rough classifier. The local feature network aims to learn the local semantic features of the color so as to maintain the contrast among different color blocks, while the

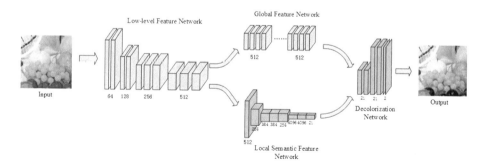

Fig. 5.1 Overview of the contrast preserving image decolorization method combining global features and local semantic features [26]. This framework is composed of four parts: a low-level features network, a local semantic feature network, a global feature network, and a decolorization network. The four components are tightly coupled so as to learn a complex color-to-gray mapping. The low-level features network uses four groups of convolution layers to extract low-level features from the input image. With the FCN (Fully Convolutional Networks) structure, the local semantic feature network acquires instance semantic information with semantic tags of an image, such as a dog and an airplane. The global feature network serves to produce global image features by processing low-level features with several convolution layers. Finally, the decolorization network with the Euclidean loss outputs the resulting grayscale image

rough classifier classifies three types of exposure states: low-exposure, normal-exposure, and overexposure features of an image. Figure 5.2 shows the ability of this method to treat images with different exposures.

(a)   (b)   (c)

Fig. 5.2 A comparison of the results with and without the exposure feature network [27]. From left to right are input images (**a**), results without (**b**) and with (**c**) the exposure feature network. From top to bottom row represent low-exposure, overexposure, and normal-exposure

5.2   Video Decolorization

As for video decolorization, people mostly extend image decolorization methods to process video frames, which would easily lead to the flicker phenomenon due to the spatiotemporal inconsistency. Video decolorization should take into account both the contrast preservation of each video frame and the temporal consistency between video frames.

Since the method of Smith et al. [6] can preserve consistency by avoiding changes in color ordering, they extended their two-step image grayscale transformation method to treat video decolorization. Owing to the ability to maintain consistency over varying palettes, Ancuti et al. [13] applied their fusion-based decolorization technique for video cases. Given a video, Ancuti et al. [14] searched in the entire sequence for the color palette that appears in each image (mostly identified with the static background). In this way, they extend their saliency-guided decolorization approach to video decolorization. For a video with a relatively constant color palette, they computed a single offset angle value for the middle frame in a video.

Song et al. [28] proposed a real-time video decolorization method using bilateral filtering. Considering that the human visual system is more sensitive to luminance than the chromaticity values, they recover the color contrast/detail loss in the luminance. They represent the loss as a residual image by the bilateral filter. The resulting grayscale image is a sum of the residual image and the luminance of the original color image. Since the residual image is robust to temporal variations, this method can preserve the temporal coherence between video frames. Moreover, as the kernel of the bilateral filter can be set as large as the input image, this method is efficient and can run in real time on a 3.4 GHz i7 CPU.

Tao et al. [29, 30] defined decolorization proximity to measure the similarity of adjacent frames and presented a temporal-coherent video decolorization method using proximity optimization. They then, respectively, treat frames with low, medium, and high proximities in order to better preserve the quality of these three types of frames. Finally, with a decolorization Gaussian mixture model (DC-GMM), they classify the frames and assign appropriate decolorization strategies to them via their corresponding decolorization proximity.

Most of the existing video decolorization methods directly apply image decolorization algorithms to treat video frames, which would easily cause temporal inconsistency and flicker phenomenon. Moreover, there may be similar local content features between video frames, which can be used to avoid redundant information. To this end, Liu and Zhang [31] introduced deep learning into the field of video decolorization by using CNN and a long short-term memory neural network. To the best of our knowledge, this is among the first attempts to perform video decolorization using deep learning techniques. A local semantic content encoder was designed to learn the same local content of a video. Here, the local semantic features were further refined by a temporal feature controller via a bi-directional recurrent neural network with long short-term memory units. Figure 5.3 shows an overview of this method.

Table 5.1 summarizes the image and video decolorization methods.

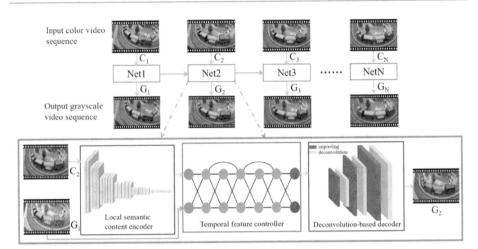

Fig. 5.3 The framework of the video decolorization method based on the CNN and LSTM neural network [31]. Given a video sequence $C_t | t = 1, 2, 3, ..., N$, it is processed into sequence images. Then the local semantic content encoder extracts deep features of these sequence images, adjusts the scale of the feature maps, and inputs them to the temporal features controller. After the output feature maps are fed into the deconvolution-based decoder, the resulting grayscale video sequence $G_t | t = 1, 2, 3, ..., N$ is produced

Table 5.1 Image and video decolorization methods

Methods	References
Local decolorization	Bala and Eschbach [4], Neumann et al. [5]
	Smith et al. [6], Zhang and Liu [8]
Global decolorization	Grundland and Dodgson [10]
	Gooch et al. [9], Kuk et al. [11]
	Kim et al. [12], Ancuti et al. [13]
	Lu et al. [7], Lu et al. [19]
	Song et al. [21], Lu et al. [20]
	Liu et al. [15], Ji et al. [23]
	Liu et al. [16], Sowmya et al. [22]
	Song et al. [18]
Deep decolorization	Hou et al. [25], Zhang and Liu [26]
	Liu and Zhang [27]
Video decolorization	Smith et al. [6], Ancuti et al. [13]
	Song et al. [28], Tao et al. [29]
	Tao et al. [30], Liu and Zhang [31]

References

1. R.S. Hunter, Photoelectric color difference meter. J. Opt. Soc. Am. **48**(12), 985–993 (1958)
2. G. Wyszecki, W.S. Stiles, Color science: concepts and methods, quantitative data and formulas. Phys. Today **21**(6), 83–84 (1968)
3. Y. Nayatani, Simple estimation methods for the Helmholtz-Kohlrausch effect. Color Res. Appl. **22**(6), 385–401 (2015)
4. R. Bala, R. Eschbach, Spatial color-to-grayscale transform preserving chrominance edge information, in *Proceedings of the 12th Color Imaging Conference*, pp. 82–86 (2004)
5. L. Neumann, M. Čadík, A. Nemcsics, An efficient perception-based adaptive color to gray transformation, in *Proceedings of Eurographics Conference on Computational Aesthetics in Graphics, Visualization and Imaging*, pp. 73–80 (2007)
6. K. Smith, P.E. Landes, J. Thollot, M. Karol, Apparent greyscale: a simple and fast conversion to perceptually accurate images and video. Comput. Graph. Forum **27**(2), 193–200 (2008)
7. C.W. Lu, X. Li, J.Y. Jia, Contrast preserving decolorization, in *Proceedings of IEEE International Conference on Computational Photography (ICCP)*, pp. 1–7 (2012)
8. H. Zhang, S. Liu, Efficient decolorization via perceptual group difference enhancement, in *Proceedings of the International Conference on Image and Graphics (ICIG)*, pp. 560–569 (2017)
9. A.A. Gooch, S.C. Olsen, J. Tumblin, B. Gooch, Color2Gray: salience-preserving color removal, in *Proceedings of ACM SIGGRAPH*, pp. 634–639 (2005)
10. M. Grundland, N.A. Dodgson, Decolorize: fast, contrast enhancing, color to grayscale conversion. Pattern Recogn. **40**(11), 2891–2896 (2007)
11. J.G. Kuk, J.H. Ahn, N.I. Cho, A color to grayscale conversion considering local and global contrast, in *Proceedings of Asian Conference on Computer Vision (ACCV)*, pp. 513–524 (2010)
12. Y.J. Kim, C.H. Jang, J.L. Demouth, S.Y. Lee, Robust color-to-gray via nonlinear global mapping, in *Proceedings of ACM SIGGRAPH Asia*, pp. 161:1–161:4 (2009)
13. C.O. Ancuti, C. Ancuti, C. Hermans, P. Bekaert, Image and video decolorization by fusion, in *Proceedings of the Asian Conference on Computer Vision (ACCV)*, pp. 79–92 (2010)
14. C.O. Ancuti, C. Ancuti, P. Bekaert, Enhancing by saliency-guided decolorization, in *Proceedings of IEEE Conference on Computer Vision and Pattern Recognition (CVPR)*, pp. 257–264 (2011)
15. Q.G. Liu, J.J. Xiong, L. Zhu, M.H. Zhang, Y.H. Wang, GcsDecolor: gradient correlation similarity for efficient contrast preserving decolorization. IEEE Trans. Image Process. **24**(9), 2889–2904 (2015)
16. Q.G. Liu, J.J. Xiong, L. Zhu, M.H. Zhang, Y.H. Wang, Extended RGB2Gray conversion model for efficient contrast preserving decolorization. Multimed. Tools Appl. **76**(12), 1–20 (2016)
17. W.F. Xue, L. Zhang, X.Q. Mou, A.C. Alan, Gradient magnitude similarity deviation: a highly efficient perceptual image quality index. IEEE Trans. Image Process. **23**(2), 684–695 (2014)
18. M.L. Song, D.C. Tao, C. Chen, X.L. Li, C.W. Chen, Color to gray: visual cue preservation. IEEE Trans. Pattern Anal. Mach. Intell. **32**(9), 1537–1552 (2010)
19. C.W. Lu, L. Xu, J.Y. Jia, Real-time contrast preserving decolorization, in *Proceedings of SIGGRAPH Asia 2012 Technical Briefs*, pp. 1–4 (2012)
20. C.W. Lu, L. Xu, J.Y. Jia, Contrast preserving decolorization with perception-based quality metrics. Int. J. Comput. Vis. **110**(2), 222–239 (2014)
21. Y.B. Song, L.C. Bao, X.B. Xu, Q.X. Yang, Decolorization:is rgb2gray() out?, in *Proceedings of ACM SIGGRAPH Asia Technical Briefs*, pp. 1–4 (2013)
22. V. Sowmya, D. Govind, K.P. Soman, Significance of incorporating chrominance information for effective color-to-grayscale image conversion. Signal Image Video Process **11**(1), 1–8 (2016)

23. Z.P. Ji, M.E. Fang, Y.G. Wang, W.Y. Ma, Efficient decolorization preserving dominant distinc-
    tions. Vis. Comput. **32**(12), 1–11 (2016)
24. Z. Lin, W. Zhang, X. Tang, Learning partial differential equations for computer vision, Microsoft,
    Technical report MSR-TR-2008-189 (2008)
25. X.X. Hou, J. Duan, G.P. Qiu, Deep feature consistent deep image transformations: downscaling,
    decolorization and HDR tone mapping (2017). arXiv:1707.09482
26. X. Zhang, S. Liu, Contrast preserving image decolorization combining global features and local
    semantic features. Vis. Comput. **34**(6), 1099–1108 (2018)
27. S. Liu, X. Zhang, Image decolorization combining local features and exposure features. IEEE
    Trans. Multimed. **21**(10), 2461–2472 (2019)
28. Y.B Song, L.C. Bao, Q.X. Yang, Real-time video decolorization using bilateral filtering, in
    Proceedings of the IEEE Winter Conference on Applications of Computer Vision (WACV), pp.
    159–166 (2014)
29. Y. Tao, Y. Shen, B. Sheng, P. Li, E. Wu, Temporal coherent video decolorization using proximity
    optimization, in *Proceedings of the 33rd Computer Graphics International Conference*, pp. 41–
    44 (2016)
30. Y. Tao, Y. Shen, B. Sheng, P. Li, R.H. Lau, Video decolorization using visual proximity coherence
    optimization. IEEE Trans. Cybern. **48**(5), 1406–1419 (2017)
31. S. Liu, H. Wang, X. Zhang, Video decolorization based on CNN and LSTM neural network, in
    ACM Transactions on Multimedia Computing, Communications, and Applications, vol. 17, no.
    3, pp. 88:1–88:18 (2021)

Style Transfer

As an extension of color transfer, style transfer refers to rendering the content of a target image or video in the style of an artist with either a style sample or a set of images through a style transfer model (Fig. 6.1). As an emerging field, the study of style transfer has attracted the attention of a large number of researchers. After decades of development, it has become a highly interdisciplinary research with a variety of artistic expression styles that can be achieved, such as pencil painting, oil painting, watercolor painting, cartoon painting, and other artistic effects.

Style transfer is a derivative of early texture synthesis. The early texture synthesis algorithm only processes and learns low-level features. The high learning ability of deep learning allows style transfer to be more flexible and efficient. Therefore, it has been used in many applications. For example, on social networking websites, users can share their art works, such as PRISMA, Ostagram, and Deep Forger. In addition, it can be used as a user-aided creation tool, which helps painters create specific style art works more conveniently. It can also be applied in fashion design, film, animation, and game creation.

In the research of style transfer, how to ensure the efficiency, flexibility, and quality of style transfer has been challenges that researchers have been trying to solve. High efficiency can help users obtain a processing result in real time; a flexible style method can help users produce a desired style result without having to retrain the model, which is time-consuming and memory-consuming. Quality-aware style transfer methods attempt to produce high-quality transfer results for different styles.

6.1 Image Style Transfer

Gatys et al. [1] proposed an image style migration approach based on a convolutional neural network (CNN). They demonstrate that CNN can not only encode image content information, but also can encode image style information by constructing the Gram matrix. Therefore,

© The Author(s), under exclusive license to Springer Nature Switzerland AG 2023
S. Liu, *Image and Video Color Editing*, Synthesis Lectures on Visual Computing: Computer Graphics, Animation, Computational Photography and Imaging,
https://doi.org/10.1007/978-3-031-26030-8_6

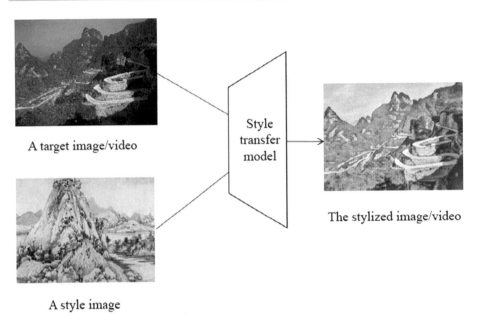

A target image/video

A style image

Fig. 6.1 An illustration of image or video style transfer

CNN can be used to separate the abstract feature representation of image content and style, and the effect of art transfer can be completed by processing these feature representations independently.

Later, researchers have successively improved this method, such as optimization-based method [2, 3]. This type of method depends on iterative optimization, and though impressive results can be produced, it is inefficient and not applicable in practical applications. Some researchers learned neural networks to generate similar outputs in one quick forward pass.

6.1.1   Efficiency-Aware Style Transfer

Johnson et al. [4] proposed a pre-trained feed-forward neural network to minimize the same objective to replace the optimization process in [1], which is called fast style transfer for the system overview. This method uses the perception loss function to train feed-forward networks for a specific style. It produces comparable results; however, it is three orders of magnitude faster than [1].

Li and Wand [5] and Ulyanov et al. [3] employed a similar network architecture for style transfer. These feedforward transmission strategies can be three orders of magnitude faster than the optimization-based methods. However, since these methods use the batch normalization for rough feature processing, which may result in poor transfer quality. The above fast stylization method via generator networks is promising, however suffering from a less satisfactory visual quality and diversity compared to generation-by-optimization. Ulyanov et al. [6] improved this type of method from two aspects: (a) using an instance normalization (IN) module to replace batch normalization (BN) to improve the quality of stylization, and (b) encouraging generators to sample unbiasedly from the Julesz texture ensemble to improve the diversity of the stylization. Experiments showed that this method can produce stylization results much closer to those of generation-via-optimization at a faster speed.

Deng et al. [7] presented a multi-adaptation image style transfer (MAST) approach, which can adaptively combine the content and style features by disentangling them. Wang et al. [8] proposed multi-modal transfer, a multi-modal convolutional neural network for style transfer. This is a hierarchical training scheme for fast style transfer which is able to learn artistic style cues at multiple scales (e.g., color, texture structure, and exquisite brushwork). In this method, the feature representation of color channel and brightness channel is added, and the style processing is carried out in the way of multi-scale classification. This method can effectively solve the problem of texture scale adaptation, producing more visually appealing results for high-resolution images. Zhang and Dana [9] constructed a Multi-style Generative Network (MSG-Net) to achieve real-time style transfer. This method introduces the ComMtch layer to learn to match the second-order statistical features with the target style, which is compatible with some existing content style interpolation methods, color preservation methods, spatial control methods, and stroke size control methods. In this method, an upsampled convolution is specially designed to reduce the checkerboard artifacts due to fractionally strided convolution.

These methods perform pre-training based on a single style, i.e., the user needs to generate a corresponding style model for each type of style. Although the efficiency and quality of this kind of method are fine, the flexibility is limited and unsuited to practical applications.

6.1.2   Flexibility-Aware Style Transfer

The above feed-forward-based methods are efficient but at the cost of sacrificing generality. These methods usually suffer from a lack of generality (e.g., constructing one network per style) and diversity (e.g., generating visually similar results). To this end, Li et al. [10] proposed a deep generative feed-forward network, allowing the efficient synthesis of multiple textures within a single network and interpolation between them. They designed a diversity loss to avoid being trapped in a degraded solution and produce diverse texture samples. Moreover, an incremental learning algorithm is used to achieve better style transfer quality. This method can synthesize up to 300 textures and transfer 16 styles to avoid the problem of limited types of style.

Dumoulin et al. [11] built a single, scalable deep network to capture the artistic style of a diversity of paintings. This model allows the user to synthesize new painting styles by arbitrarily combining the styles learned from individual paintings. It has great generalization across diverse artistic styles by reducing a painting to a point in an embedding space. In this method, the conditional normalization is employed to learn the different parameters of each style in a group of styles. The network can generate completely different styles of images with the same convolution parameters, which realizes the transfer of 32 groups of styles. Without changing the conditional normalization layer, Ghiasi et al. [12] added a style prediction network to directly learn a mapping between a style image and style parameters, which balances between limited flexibility and efficiency.

Huang and Belongie [13] proposed an adaptive instance normalization layer to realize efficient and arbitrary image style transfer. Different from the conditional normalization layer, the adaptive instance normalization layer processes any style, and there is no limit to the number of fixed styles. This method solves the problem of high efficiency and arbitrary style conversion, but it may generate less satisfactory quality. To this end, Xu et al. [14] added mask and discriminator layers on the basis of adaptive instance normalization layer and introduced the discriminator of GAN to distinguish the generated image from the real image, so as to make the structure of the transferred style result clear. In addition, the results produced by the model can fit for various style fields.

The above methods rely too much on the VGG network of encoding and decoding, costing a relatively large memory. Therefore, Shen et al. [15] proposed to utilize meta network for style transfer, which not only further improves the efficiency, but also occupies less memory. This method has great potential in applying to mobile devices.

6.1.3   Quality-Aware Style Transfer

Although the visual effect and efficiency of style transfer have been greatly improved, the existing methods cannot coordinate the visual distribution of content image and style image in different spaces, or present different levels of detail through different strokes.

Gatys et al. [16] controlled image styles from three perspectives: spatial location, color information, and spatial scale. This strategy can thereby achieve high-resolution stylization effects and reduces general failure cases (e.g., transferring ground textures to sky areas). It is a challenge to control the stroke size when transferring a photo to an artistic style. Jing et al. [17] proposed a stroke controllable style transfer method to realize continuous and spatial stroke size control by considering the receptive field and the style image scales. Distinct stroke sizes in different spatial regions can be generated in the output stylized results.

Although the visual effects can be greatly improved, it fails to coordinate the spatial distribution of visual attention between the content image and the output. To this end, Yao et al. [18] proposed an attention-aware multi-stroke style transfer method. A self-attention mechanism is introduced to an autoencoder network to generate an attention map of a content image denoting the critical characteristics and long-range region relations. Moreover, a multi-scale style swap is employed to produce multiple feature maps reflecting different stroke patterns. Experiments demonstrated that this method can produce convincing results in terms of both stylization effects and visual consistency. Park and Lee [19] proposed an arbitrary style transfer method via style-attentional networks (SANets). Given a content image and a style image, arbitrary style transfer aims to create a new image having the style patterns while preserving the content structure. The SANet incorporates the local style patterns based on the semantic spatial distribution of the content image. With a special loss function and multi-level feature embeddings, the SANet is able to maintain the content structure and respect the style patterns. Experiments showed that this approach can generate high-quality stylized results, reaching a good balance between preserving the global and local style patterns and maintaining the content structure. When there are multiple objects at different depths in the content image, the produced stylized results by previous methods suffer from destroyed structures due to inaccurate depth information [20]. Liu et al. [21] proposed a depth-aware neural style transfer approach that introduces depth preservation as an additional loss to estimate the depth differences between the stylized image and the content target image. Although Liu et al. [21] can improve the style transfer rendering quality, it needs to train an individual neural network for each style. Cheng et al. [22] further presented a structure-aware image style transfer framework, which improves the visual quality owing to both a global structure extraction network and a local structure refinement network. Kozlovtsev and Kitov [23] extended the AdaIN method [13] by preserving the depth information from the content image. Kotovenko et al. [24] proposed a content transformation block for image style transfer, using similar content in the target image and style samples to learn how style alters content details.

Deng et al. [25] introduced Transformers into image style transfer and proposed StyTr2Image. Sanakoyeu et al. [26] introduced a style-aware content loss to achieve real-time, high-resolution stylization of images and videos. A new measure is specially designed to quantitatively estimate the quality of a stylized image.

6.2   Video Style Transfer

It is very tedious to manually perform image style transfer. Therefore, it is beyond imagination to stylize a video manually. Given a source style image (e.g., a van Gogh painting) and a target video, a video style transfer technique aims to automatically transfer the style of the source image to the whole target video sequence. Recently, researchers have developed some video style transfer techniques to alleviate this painstaking task.

In 2016, Ruder et al. [27] extended previous image style transfer methods for video cases in an optimization framework. This method includes ingredients specific to video style transfer, namely a new initialization, a loss function for short-term temporal consistency, a loss function for long-term consistency, and a multi-pass approach. This method is able to handle cases with large motion and strong occlusion in a video. Later, they [28] extended the above method for artistic style transfer for both videos and spherical images by regarding video style transfer as a learning problem. Rather than solving an optimization minimization problem, they learn a style transfer function from an input image to its stylized output in a much faster manner. This method can prevent the propagation of errors by iterating the network times during training. Anderson et al. [29] proposed DeepMovie, a movie style transfer method by combining optical flow and deep neural networks. In this method, in order to make the textures move with the objects in a video frame, they use the optical flow to initialize the style transfer optimization.

It will introduce flickering phenomena if we perform video style transfer frame by frame. Chen et al. [30] presented a coherent online video style transfer method, which can produce a temporally coherent stylized video in near real time. Their feed-forward network consists of a flow sub-network and a mask sub-network, which are integrated into an intermediate layer of a pre-trained stylization network (e.g., [4] and [2]).

Huang et al. [31] proposed a real-time neural style transfer approach for videos. Their network model is comprised of a stylizing network and a loss network. The former takes one frame as input and outputs the corresponding stylized result. The latter is pre-trained on the ImageNet. It takes in charge of learning features of the stylized results and calculates the losses for training the stylizing network. The two subnetworks are highly coupled for efficiency. In order to balance the perceptual style quality and temporal consistency, Gao et al. [32] proposed ReCoNet, a real-time coherent video style transfer network. A luminance warping constraint is incorporated in the output-level temporal loss by taking care of luminance changes of traceable pixels in the original video. This strategy improves the stability in illumination-aware regions and enhances the overall temporal consistency. They also designed a feature-map-level temporal loss to restrict the changes in high-level features of the same object in neighboring frames. Gao et al. [33] further proposed a fast video multi-style transfer (VMST) method, enabling a fast and multi-style video transfer in a single network. In VMST, a multi-instance nor- malization block (MIN-Block) is used to learn different style examples (i.e., 120 examples) and two ConvLSTM modules are employed to ensure both the short- and long-term temporal consistency between video frames.

Frigo et al. [34] extended the patch-based style transfer method to video cases. They leverage the optical flow technique to track video frames. A spatially and temporally consistent patch synthesis strategy is used to alleviate the inconsistencies during tracking.

Gupta et al. [35] improved the stability of the neural style transfer method by showing that the trace of the Gram matrix representing style is inversely related to the stability of the neural style transfer method. By designing a temporal consistency loss, they employed a recurrent convolutional network (R-CNN) to achieve real-time, stable video style transfer method. Given the previous stylized frame and the target frame, this method aims to stylize. This is a R-CNN by using the output at each time step as input for the next cycle. Note that the style and content losses are employed to ensure similarity and the at each time step enforce style and content losses to ensure similarity with the input frame and temporal consistency loss is used for achieving stability between frames. Experiments showed that this method gains 1000× speed improvement without sacrificing the video transfer quality.

Deng et al. [36] presented a Multi-Channel Correlation network (MCCNet) for efficient video style transfer. To make the generated stylized results preserve clear content structures, they analyze the multi-channel correlation between content and style features. This method is suited for dealing with complex light conditions owing to an illumination loss to ensure a stable style transfer process.

Due to missing texture information, blurred boundaries and distortion usually occur in the stylized image. It is still a challenge to preserve the hierarchical structure of the content image in video style. To this end, Liu and Zhu [37] proposed a structure-guided arbitrary style transfer method for both artistic images and videos. In this method, in order to diversify the stylized results with levels of details, a refined network [38] is embedded into an autoencoder framework to guide the style transfer. Both a global content loss and a local region structure loss are designed to train the model. In the video cases, a temporal consistency loss and a

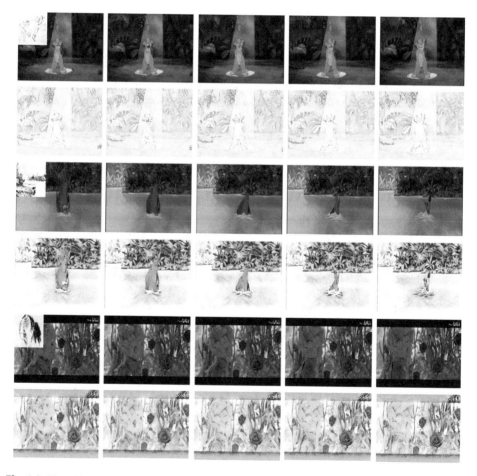

Fig. 6.2 The video style transfer results by Liu and Zhu's method [37]

Table 6.1 Image and video style transfer methods

Methods	References
Efficiency-aware style transfer	Gatys et al. [1], Johnson et al. [4], Li and Wand [5], Ulyanov et al. [3], Ulyanov et al. [6], Wang et al. [8], Zhang and Dana [9], Deng et al. [7]
Flexibility-aware style transfer	Dumoulin et al. [11], Ghiasi et al. [12], Huang and Belongie [13]. Li et al. [10], Shen et al. [15], Xu et al. [14]
Quality-aware style transfer	Gatys et al. [16], Liu et al. [21], Jing et al. [17], Kozlovtsev and Kitov [23], Park and Lee [19], Yao et al. [18], Cheng et al. [22]
Video style transfer	Anderson et al. [29], Ruder et al. [27], Chen et al. [30], Gupta et al. [35], Huang et al. [31], Gao et al. [32], Ruder et al. [28], Frigo et al. [34], Gao et al. [33], Deng et al. [36], Liu and Zhu [37]

cycle-temporal loss are further introduced to reduce temporal incoherence and motion blur. This method may fail to deal with a target video of which the contents contain large motions. Figure 6.2 style transfer results by Liu and Zhu's method. Note that three different comic clips were tested with different style patterns.

Table 6.1 summarizes the image and video style transfer methods.

References

1. L.A. Gatys, A.S. Ecker, M. Bethge, Image style transfer using convolutional neural networks, in *Proceedings of IEEE Conference on Computer Vision and Pattern Recognition (CVPR)* (2016), pp. 2414–2423
2. D. Chen, L. Yuan, J. Liao, N. Yu, G. Hua, StyleBank: an explicit representation for neural image style transfer, in *IEEE Conference on Computer Vision and Pattern Recognition (CVPR)* (2017), pp. 2770–2779
3. D. Ulyanov, V. Lebedev, A. Vedaldi, V.S. Lempitsky, Texture networks: Feed-forward synthesis of textures and stylized images, in *Proceedings of the 33rd International Conference on International Conference on Machine Learning (ICML)* (2016), pp. 1349-1357
4. J. Johnson, A. Alahi, F.F. Li, Perceptual losses for real-time style transfer and super-resolution, in *Proceedings of European Conference on Computer Vision (ECCV)* (2016), pp. 694–711
5. C. Li, M. Wand, Precomputed real-time texture synthesis with markovian generative adversarial networks, in *Proceedings of European Conference on Computer Vision (ECCV)* (2016), pp. 702–716
6. D. Ulyanov, A. Vedaldi, V. Lempitsky, Improved texture networks: maximizing quality and diversity in feed-forward stylization and texture synthesis, in *Proceedings of IEEE Conference on Computer Vision and Pattern Recognition* (2017), pp. 4105–4113
7. Y. Deng, F. Tang, W. Dong, W. Sun, F. Huang, C. Xu, Arbitrary style transfer via multi-adaptation network, in *Proceedings of ACM Multimedia* (2020), pp. 2719–2727

8. X. Wang, G. Oxholm, D. Zhang, Y.-F. Wang, Multimodal transfer: a hierarchical deep convolutional neural network for fast artistic style transfer, in *Proceedings of IEEE Conference on Computer Vision and Pattern Recognition (CVPR)* (2017), pp. 7178–7186

9. H. Zhang, K. Dana, Multi-style generative network for real-time transfer, in *Proceedings of European Conference on Computer Vision (ECCV) Workshops* (2018), pp. 349–365

10. Y. Li, C. Fang, J. Yang, Z. Wang, X. Lu, M.-H. Yang, Diversified texture synthesis with feed-forward networks, in *Proceedings of IEEE Conference on Computer Vision and Pattern Recognition (CVPR)* (2017), pp. 266–274

11. V. Dumoulin, J. Shlens, M. Kudlur, A learned representation for artistic style, in *Proceedings of International Conference on Learning Representations (ICLR)* (2017)

12. G. Ghiasi, H. Lee, M. Kudlur, et al. Exploring the structure of a real-time, arbitrary neural artistic stylization network, in *Proceedings of British Machine Vision Conference* (2017)

13. X. Huang, S. Belongie, Arbitrary style transfer in real-time with adaptive instance normalization, in *Proceedings of IEEE International Conference on Computer Vision (ICCV)* (2017), pp. 1510–1519

14. Z. Xu, M.J. Wilber, C. Fang C, et al., Learning from multi-domain artistic images for arbitrary style transfer, in *Proceedings of Eurographics Expressive Symposium* (2019), pp. 21–31

15. F. Shen, S. Yan, G. Zeng, Neural style transfer via meta networks, in *Proceedings of IEEE Conference on Computer Vision and Pattern Recognition (CVPR)* (2018), pp. 8061–8069

16. L. Gatys, A. Ecker, M. Bethge, A. Hertzmann, E. Shechtman, Controlling perceptual factors in neural style transfer, in *Proceedings of IEEE Conference on Computer Vision and Pattern Recognition (CVPR)*, (2017), pp. 3730–3738

17. Y. Jing, Y. Liu, Y. Yang, et al., Stroke controllable fast style transfer with adaptive receptive fields, in *Proceedings of European Conference on Computer Vision (ECCV)* (2018), pp. 244–260

18. Y. Yao, J. Ren, X. Xie, W. Liu, Y. Liu, J. Wang, Attention-aware multi-stroke style transfer, in *Proceedings of IEEE Conference on Computer Vision and Pattern Recognition (CVPR)* (2019), pp. 1467–1475

19. D. Park, K. Lee, Arbitrary style transfer with style-attentional network, in *Proceedings of IEEE Conference on Computer Vision and Pattern Recognition (CVPR)* (2019), pp. 5880–5888

20. W. Chen, Z. Fu, D. Yang, et al., Single-image depth perception in the wild, in *Proceedings of Advances in Neural Information Processing Systems (NIPS)* (2016), pp. 730–738

21. X. Liu, M. Cheng, Y. Lai, et al., Depth-aware neural style transfer, in *Proceedings of International Symposium on Non-Photorealistic Animation and Rendering* (2017), pp. 1–10

22. M. Cheng, X. Liu, J. Wang, S. Lu, Y. Lai, P.L. Rosin, Structure-preserving neural style transfer. IEEE Trans. Image Process. **29**, 909–920 (2020)

23. K. Kozlovtsev, V. Kitov, Depth-preserving real-time arbitrary style transfer (2019). arXiv:1906.01123

24. D. Kotovenko, A. Sanakoyeu, P. Ma, et al., A content transformation block for image style transfer, in *Proceedings of IEEE Conference on Computer Vision and Pattern Recognition (CVPR)* (2019), pp. 10032–10041

25. Y. Deng, F. Tang, W. Dong, C. Ma, X. Pan, L. Wang, C. Xu, StyTr2Image style transfer with transformers, in *Proceedings of IEEE/CVF Conference on Computer Vision and Pattern Recognition (CVPR)* (2022)

26. A. Sanakoyeu, D. Kotovenko, S. Lang, B. Ommer, A style-aware content loss for realtime HD style transfer, in *Proceedings of European Conference on Computer Vision (ECCV)* (2018), pp. 715–731

27. M. Ruder, A. Dosovitskiy, T. Brox, Artistic style transfer for videos, in *Proceedings of German Conference on Pattern Recognition* (2016), pp. 26–36

28. M. Ruder, A. Dosovitskiy, T. Brox, Artistic style transfer for videos and spherical images. Int. J. Comput. Vis. **126**(11), 1199–1219 (2018)

29. A. Anderson, C. Berg, D. Mossing, et al., DeepMovie: using optical flow and deep neural networks to stylize movies (2016). arXiv:1409.1556

30. D. Chen, J. Liao, L. Yuan, et al., Coherent online video style transfer, in *Proceedings of IEEE International Conference on Computer Vision (ICCV)* (2017), pp. 1114–1123

31. H. Huang, H. Wang, W. Luo, L. Ma, W. Jiang, X. Zhu, Z. Li, W. Liu, Real-time neural style transfer for videos, in *Proceedings of IEEE Conference on Computer Vision and Pattern Recognition (CVPR)* (2017), pp. 7044–7052

32. C. Gao, D. Gu, F. Zhang, Y. Yu, ReCoNet: real-time coherent video style transfer network, in *Proceedings of Asian Conference on Computer Vision (ACCV)* (2018), pp. 637–653

33. W. Gao, Y. Li, Y. Yin, M.-H. Yang, Fast video multi-style transfer, in *Proceedings of IEEE Winter Conference on Applications of Computer Vision (WACV)* (2020), pp. 3211–3219

34. O. Frigo, N. Sabater, J. Delon, P. Hellier, Video style transfer by consistent adaptive patch sampling. Vis. Comput. **35**(3), 429–443 (2019)

35. A. Gupta, J. Johnson, A. Alahi, et al., Characterizing and improving stability in neural style transfer, in *Proceedings of IEEE International Conference on Computer Vision (ICCV)* (2017), pp. 4087–4096

36. Y. Deng, F. Tang, W. Dong, H. Huang, C. Ma, C. Xu, Style transfer via multi-channel correlation, in *Proceedings of The Thirty-Fifth AAAI Conference on Artificial Intelligence (AAAI)* (2021), pp. 1210–1217

37. S. Liu, T. Zhu, Structure-guided arbitrary style transfer for artistic image and video. IEEE Trans. Multimed. **24**, 1299–1312 (2022)

38. T. Zhu, S. Liu, Detail-preserving arbitrary style transfer, in *Proceedings of the IEEE International Conference on Multimedia and Expo (ICME)* (2020), pp. 1–6

Low-Exposure Image and Video Enhancement

Low-exposure image and video enhancement aims to process color and details of an underexposed image or video to improve its appearance. With the rapid development of digital photography technology, the acquisition of digital images and videos becomes more and more convenient. However, in daily life, due to low lighting, equipment configuration, and other reasons, the images and videos obtained often suffer from being underexposed and show dark scenes. Low-exposure images and videos have very low visibility, and the details in the image and video scenes cannot be displayed well, which greatly affects the aesthetics and has an influence on the downstream image and video processing. Especially in the public security management, the monitoring system usually suffers from capturing low-exposure night vision video, which seriously affects the public transport monitoring and the monitoring of criminal activities. In the downstream processing of images and videos, the processing (e.g., tone adjustment and pattern recognition) would be greatly influenced by insufficient exposure and will be affected. Therefore, these images and videos must be enhanced by effective methods to make their exposure uniform and display enough details.

7.1 Low-Exposure Image Enhancement

The goal of underexposed image enhancement is to expand the dynamic range of the image so that the details in the image can be better displayed. At present, the image enhancement methods include histogram equalization, tone mapping, the Retinex theory-based methods, image fusion, deep learning-based methods, etc.

© The Author(s), under exclusive license to Springer Nature Switzerland AG 2023
S. Liu, *Image and Video Color Editing*, Synthesis Lectures on Visual Computing: Computer Graphics, Animation, Computational Photography and Imaging,
https://doi.org/10.1007/978-3-031-26030-8_7

7.1.1 Histogram Equalization

When expanding the dynamic range of a low-exposure image, the histogram equalization [1, 2] is a common method. This method transforms the histogram of the original image into a uniformly distributed histogram, in order to increase the contrast.

Trahanisa et al. [3] proposed an image equalization method using histograms of three color channels, replacing the previous method of using one-dimensional histogram equalization for each channel of R, G, and B. They defined a uniform probability density function PDF (Probability Density Function) and a cumulative distribution function CDF (Cumulative Distribution Function) in the RGB color space, taking these two functions as ideal histogram functions. Although the uniform probability density function PDF in the gray space can significantly enhance the contrast of the image, the uniform PDF in the three-dimensional space does not represent the uniform PDF in the luminance domain. Most natural images are homogenized using this method, which will result in a more concentrated pixel distribution with higher brightness. In 2007, Sim et al. [4] proposed that the goal of dynamic histogram equalization is to expand the contrast of the original image while maintaining the details in the original image. The dynamic histogram equalization method removes the minimum value of image brightness and assigns a new dynamic range to each sub-histogram. To ensure accurate primary partitioning, they further refined the histogram by re-partitioning the test. However, their dynamic histogram equalization method still cannot maintain the uniform brightness distribution of the image and still has the problem of brightness saturation.

Later, Kim and Chung [5] proposed a method called recursive separation and weighted histogram equalization RSWHE (Recursion Separation and Weighted Histogram Equalization) to perform weighted segmentation of image histograms. They proposed two different weighted histogram equalization segmentation methods: average segmentation and median segmentation. It can be seen from the experimental results that, compared with the median segmentation method, the average segmentation method can better maintain the brightness smoothness. Global histogram equalization may cause uneven brightness distribution. Based on this consideration, Brajovic et al. [6] proposed a method based on pixel-by-pixel exposure matching for the histogram. However, this method does not take into account the regional differences of the scene objects, and the calculated exposure values are not suitable for each region. It is easy to have different exposures in the same region, and it lacks accuracy in maintaining the contrast of different objects.

7.1.2 Tone Mapping

Tone mapping is also a common image enhancement method [7–10]. By adjusting the tone curve, a more suitable tone curve can be obtained to enhance a given image or video. However, this method needs to calculate different tone curves for different images or videos and cannot guarantee the overall contrast. There are two main tone mapping methods: global tone mapping and local tone mapping.

The global tone mapping method [11, 12] uniformly remaps the image intensity in the spatial domain to compress the dynamic range of the image. The global tone mapping methods are simple with fast implementation. However, this method does not take into account the local contrast and other information, which would produce unsatisfactory results.

The local tone mapping method makes different mapping changes in different spatial domains [9, 10]. Compared with the global mapping method, this method can obtain better results. Yuan et al. [13] proposed a region-based exposure correction method. First, the region-based exposure estimation is performed on the image to ensure the contrast between different regions. Then, an S-spline curve is used to adjust the hue to maintain the image details. However, this method is actually a global-based method when considering image details, which is difficult to preserve image details with unclear boundaries, and cannot be directly applied to video processing. Eilertsen et al. [14] proposed a tone mapping method that can be used for real-time processing. First, local adaptive tone curves are calculated to minimize contrast distortion, and then a fast detail enhancement method is combined to reduce the phenomenon of over-smoothing. This method can effectively reduce the noise caused by light changes in the image and ensure the details of the scene to the greatest extent. However, this method is limited by the tone mapping method itself, which cannot ensure that every frame of the video has a good result when used in video enhancement.

7.1.3   Retinex Theory-Based Methods

The Retinex algorithm is usually applied [2, 15, 16] for uneven brightness enhancement caused by illumination. The principle of Retinex theory is that the color of an object is determined by its ability to reflect light. The color displayed by an object is not affected by the non-uniformity of light, i.e., it has color constancy. The enhanced result can be obtained by removing the scene light.

Based on the methods of Land and McCann [17], most algorithms based on Retinex theory remove the light in the scene first and enhance the details by extracting reflective layers [15, 18–20]. However, although these methods can extract the details of the image, removing scene lighting destroys the naturalness of the image as a whole [23]. To this end, some methods use the algorithm of center encirclement. They use local convolution of brightness without considering the reflection coefficient [15]. But the reflection coefficient should be between 0 and 1, which means that the light reflected by the object surface cannot be brighter than the light it receives. Therefore, it is unreasonable for them to simply remove the brightness, since the light represents the overall atmosphere of the image. In order to solve this problem, Wang et al. [21] proposed that for scenes with uneven lighting, it is unsuitable to simply remove the lighting. In order to ensure the naturalness of the image scene, they first evaluate the naturalness of the scene, then separate the reflected light and light through the brightness filter, and finally process the reflected light and light separately to achieve the effect of image enhancement. However, their method does not consider the relationship between lighting in different scenes, which is prone to jitter when processing video cases. When dealing with video enhancement, it is difficult to combine the Retinex theory with the maintenance of the temporal and spatial consistency of a video.

7.1.4   Image Fusion

The image fusion can also be used to enhance low-exposure images [22–25] by combining the relevant information of multiple images from the same scene to generate a fused image, which contains the optimal brightness information and detail information in the image sequence.

Most previous exposure fusion methods perform image enhancement by the following procedure. First, they determine the contribution of each pixel or region of the input image to the final fused image according to specific quality metrics, including contrast [26, 27], texture [28], entropy [29, 30], saturation [27, 31, 32], and more; then these pixels or regions are fused to generate an image with good overall exposure. The fusion methods include pixel-based weighted average [29–31], gradient domain fusion [33, 34], pyramid-based fusion [26, 27], etc.

Mertens et al. [27] proposed a classic image fusion-based low-exposure image enhancement method. In their method, the area with a low weight value represents underexposure or over-exposure, which will not be displayed in the final merging result. Then, different exposure images of the same scene and their weight maps are fused using the Laplace pyramid method to get a great exposure result. However, this fusion-based method requires multiple input images and is not suitable for processing single-source images or videos. Song et al. [34] proposed a fusion method based on optimization, which uses an optimization method to best fuse multiple input images with different exposures. However, the weight map optimized by this method may be inaccurate, and the processed image cannot well maintain good contrast. In the above methods, multiple exposure images of different scales need to be provided for fusion. However, in most cases, only a single low-exposure image can be provided.

In order to solve the above problem, Hsieh et al. [35] proposed to take the low-exposure image after histogram equalization as one of the inputs during fusion, and then combine the original low-exposure image with a linear function to obtain the final result. On the basis of their methods, Pei et al. [36] generated two images using the input single low-exposure image, one is the image after histogram equalization and the other is the sharpened image after using the Laplacian operator, and then fused the discrete wavelet transform (DWT) coefficients of these images to obtain the fused image with higher contrast and clarity.

Different from the above methods, another fusion method does not constrain the quality of the input image before combining their information for fusion. On the contrary, this method directly restricts the quality of the output image. Raman and Chaudhuri [37] designed an energy function whose minimum value gives the optimal result of fusion. Chaudhuri and Kotwal [38] defined the final fused image as the weighted average of the input image sequence, and these weight values are the results obtained by solving the optimization problem. However, directly constraining the image itself requires adding a smooth constraint to the generated image, which may result in overly smooth results.

The inversion methods are also used to enhance images. Dong et al. [39] found that the inverted low-exposure image is similar to the image with fog, so they proposed a method of defogging the inverted low-exposure image to enhance the image effect. The final enhanced image can be obtained by defogging the inverted image and then inverting it again. However, this method cannot guarantee the contrast between objects. Later, Li et al. [40] improved this algorithm. They first segment the input image into hyperpixels, then adaptively denoise each hyperpixel block, and finally defog the inverted image to get the final result. This method depends on the high stability and accuracy of the super-pixel segmentation result. Guo [41] improved these two methods and proposed a structure-based filtering and smoothing method to enhance an underexposed scene. However, this method may fail to deal with the enhanced noise.

7.1.5 Deep Learning-Based Methods

Recently, deep learning-based low-exposure image enhancement methods emerged. Lore et al. [42] pioneered using deep learning for low-exposure image enhancement. A stacked-sparse denoising autoencoder was employed to enhance the image. Ren et al. [43] combined an encoder-decoder network and a recurrent neural network to enhance low-exposure images. The DeepUPE was presented in Wang et al. [44].

Most of the current deep learning-based methods are supervised (e.g., [43, 45–47]). In addition, there are also semi-supervised and unsupervised deep learning-based methods (e.g., [48–50]).

7.2 Low-Exposure Video Enhancement

Low-exposure video enhancement technology is also widely used in the field of video processing, especially in the processing of surveillance video. For low-exposure video enhancement, a simple way is to directly use the low-exposure image enhancement method to process the video frame by frame to enhance the whole video. However, due to the dynamic changes of the scene in the video, it is easy to have temporal and spatial inconsistencies such as jitter. This problem is more apparent under poor light shooting conditions. It is a key to keep the temporal and spatial consistency for video enhancement.

Bennett et al. [51] proposed a virtual exposure camera model, which uses an adaptive spatiotemporal consistent cumulative filter to remove noise and uses tone mapping to expand the dynamic range. Although this method can repair some videos with very low exposure, it cannot ensure the time and space consistency, and there will be some boundary bleeding in the video. Zhang et al. [52] proposed a low-exposure video color enhancement method based on perceptual-driven progressive fusion. This method calculates multiple exposure correction results for each frame of the target video according to different parameter settings, then fuses these areas with a perception-driven method to make each area of the video frame the best exposure state, and finally uses a filter that maintains details and space-time consistency to remove noise. This method can produce good results in night vision video processing. However, it is not suitable for real-time video processing, as the filtering and smoothing block processing is computationally intensive. Dong et al. [53] solved this problem and proposed a video exposure correction method based on the temporal continuity of regions, using different tone adjustment curves to enhance each region. This method solves the problems of failing to correctly estimate the best exposure rate for each region and well maintain the temporal and spatial consistency of the video. This method can ensure the temporal and spatial continuity between video frames (e.g., the objects will not change with the increase of frames) and better preserve the details of the scene. However, these methods suffer from low efficiency due to the complex region-by-region processing.

The existing image fusion methods need to input a group of image sequences with different exposure levels from high to low and then combine the optimal exposure areas in the image sequence to fuse into a well-exposed result. However, it is difficult to provide video sequences with multiple exposure levels in daily life. Liu and Zhang [54] proposed a detail-preserving underexposed video enhancement method based on a new optimal weighted multi-exposure fusion mechanism. They propose a novel multi-exposure image enhancement method that can generate a multi-exposure image sequence of each frame. However, none of these frames are good enough, as frames with high exposure have good brightness and color information, whereas sharp details are better preserved in the frames with lower exposure. Note that in the exposure images of different scales, areas with bright colors, clear details, and other characteristics are those with good image quality. The goal of exposure fusion is to combine these areas of different scales to finally form an image with good exposure. Here they determine which areas are of good quality based on three quantitative indicators (i.e., contrast, saturation, and exposure). In order to preserve details and enhance the blurred edges, they solve an energy function to compute the optimal weight of the three measurements: local contrast, saturation, and exposure. Then an optimal flow method is used to propagate color to the other frames. Figure 7.1 provides a comparison between the results by Liu and Zhang's method [54] and Mertens et al.'s method [27]. It can be observed that the proposed method can better preserve the contrast, saturation, and details.

Backlight shooting, uneven distribution of scene light sources, and other factors will cause uneven exposure of the captured video, i.e., normal exposure or over-exposure of some areas, and insufficient exposure of other areas. When using current video enhancement

(a)

(b)

(c)

Fig. 7.1 A comparison between the results by Liu and Zhang's method [54] and Mertens et al.'s method [27]. **a** the original low-exposure frames, **b** the results by Mertens et al.'s method, and **c** the results by Liu and Zhang's method

methods to process such cases, the final video is prone to non-uniform exposure because the exposure degree of different areas in the video is not considered, and the naturalness of the video cannot be maintained. Liu and Zhang [55] presented a non-uniform illumination video enhancement method based on zone system and fusion. Note that the zone system was proposed by Ansel Adams for black-and-white photography, which establishes a relationship between the scenes from the lightest and darkest scenes with divisions from pure black to pure white. They first apply the zone system for exposure evaluation. They then remap each region using a series of tone mapping curves to generate multi-exposure regions. Next, the regions are fused into a well-exposed video frame. Finally, in order to keep temporal consistency, they temporally propagate the zone regions from the former frame to the current frame using the zone system-based distance metric. Experimental results have shown that the

Fig. 7.2 A comparison between different video enhancement results. **a** an original non-uniform illumination video frame, **b** the result by Guo et al.'s method [41], **c** the result by Mertens et al.'s method [27], and **c** the results by Liu and Zhang's method [55]

enhanced video exhibits uniform exposure, keeps better naturalness, and preserves temporal consistency. Figure 7.2 shows a comparison between different video enhancement methods, including Guo et al.'s method [41], Mertens et al.'s method [27], and the proposed method by Liu and Zhang [55]. It can be seen that Guo et al.'s method and Mertens et al.'s method cannot guarantee uniform exposure. For example, the upper right corner of the table is still an over-exposed area. In contrast, Liu and Zhang's method can well ensure uniform exposure and can maintain the details of the video scene.

Recently, researchers began to use deep learning for low-exposure video enhancement (e.g., [56–58]).

References

1. A.W.M. Abdullah, M.H. Kabir, M.A.A. Dewan et al., A dynamic histogram equalization for image contrast enhancement. IEEE Trans. Consum. Electron. **53**(2), 593–600 (2007)
2. H. Malm, M. Oskarsson, E. Warrant, et al., Adaptive enhancement and noise reduction in very low light-level video, in *Proceedings of IEEE International Conference on Computer Vision* (2007), pp. 1–8

3. P.E. Trahanias, A.N. Venetsanopoulos, Color image enhancement through 3-D histogram equalization, in *Proceedings of IAPR International Conference on Pattern Recognition* (1992), pp. 545–548
4. K.S. Sim, C.P. Tso, Y.Y. Tan, Recursive sub-image histogram equalization applied to gray scale images. Pattern Recognit. Lett. **28**(10), 1209–1221 (2007)
5. M. Kim, G.C. Min, Recursively separated and weighted histogram equalization for brightness preservation and contrast enhancement. IEEE Trans. Consum. Electron. **54**(3), 1389–1397 (2008)
6. V. Brajovic, Brightness perception, dynamic range and noise: a unifying model for adaptive image sensors, in *Proceedings of IEEE Conference on Computer Vision and Pattern Recognition* (2004), pp. 189–196
7. X. Kang, S. Li, Fast multi-exposure image fusion with median filter and recursive filter. IEEE Trans. Consum. Electron. **58**(2), 626–632 (2012)
8. S. Daly, R. Mantiuk, L. Kerofsky, Display adaptive tone mapping. ACM Trans. Graph. **27**(3), 1–10 (2008)
9. M..S. Brown, Q..J. Shan, Y. Jia, Globally optimized linear windowed tone mapping. IEEE Trans. Vis. Comput. Graph. **16**(4), 663–675 (2010)
10. D. Lischinski, R. Fattal, M. Werman, Gradient domain high dynamic range compression. ACM Trans. Graph. **21**(3), 249–256 (2002)
11. K. Devlin, E. Reinhard, Dynamic range reduction inspired by photoreceptor physiology. IEEE Trans. Vis. Comput. Graph. **11**(1), 13–24 (2005)
12. F. Drago, K. Myszkowski, T. Annen, N. Chiba, Adaptive logarithmic mapping for displaying high contrast scenes. Comput. Graph. Forum **22**(3), 419–426 (2003)
13. L. Yuan, J. Sun, Automatic exposure correction of consumer photographs, in *Proceedings of European Conference on Computer Vision (ECCV)* (2012), pp. 771–785
14. G. Eilertsen, R..K. Mantiuk, J. Unger, Real-time noise-aware tone mapping. ACM Trans. Graph. **34**(6), 223–235 (2015)
15. B. Li, S. Wang, Y. Geng, Image enhancement based on Retinex and lightness decomposition, in *Proceedings of IEEE International Conference on Image Processing (ICIP)* (2011), pp. 3417–3420
16. J.H. Jang, S.D. Kim, J.B. Ra, Enhancement of optical remote sensing images by subband-decomposed multiscale Retinex with hybrid intensity transfer function. IEEE Geosci. Remote Sens. Lett. **8**(5), 983–987 (2011)
17. E.H. Land, J.J. Mccann, Lightness and Retinex theory. J. Opt. Soc. Am. **61**(1), 1–11 (1971)
18. Z.U. Rahman, D.J. Jobson, G.A. Woodell, Retinex processing for automatic image enhancement. J. Electron. Imaging **13**(1), 100–110 (2002)
19. A. Rizzi, D. Marini, L. Rovati et al., Unsupervised corrections of unknown chromatic dominants using a Brownian-path-based Retinex algorithm. J. Electron. Imaging **12**(3), 431–441 (2003)
20. B.V. Funt, F. Ciurea, J.J. Mccann, Retinex in Matlab, in *Proceedings of Color and Imaging Conference* (2000), pp. 112–121
21. S.H. Wang, J. Zheng, H.M. Hu et al., Naturalness preserved enhancement algorithm for non-uniform illumination images. IEEE Trans. Image Process. **22**(9), 3538–3548 (2013)
22. I. Adrian, R. Ramesh, J.Y. Yu, Gradient domain context enhancement for fixed cameras. Int. J. Pattern Recognit. Artif. Intell. **19**(4), 533–549 (2011)
23. R. Raskar, A. Ilie, J. Yu, Image fusion for context enhancement and video surrealism, in *Proceedings of ACM SIGGRAPH Courses* (2004), pp. 85–152
24. Y. Cai, K. Huang, T. Tan, et al., Context enhancement of nighttime surveillance by image fusion, in *Proceedings of International Conference on Pattern Recognition* (2006), pp. 980–983
25. X. Gao, S. Liu, DAFuse: a fusion for infrared and visible images based on generative adversarial network. J. Electron. Imaging **31**(4), 1–18 (2022)

26. L. Bogoni, Extending dynamic range of monochrome and color images through fusion, in *Proceedings of International Conference on Pattern Recognition* (2000), pp. 7–12
27. T. Mertens, J. Kautz, R.F. Van, Exposure fusion: a simple and practical alternative to high dynamic range photography. Comput. Graph. Forum **28**(1), 161–171 (2009)
28. S. Raman, S. Chaudhuri, Bilateral filter based compositing for variable exposure photography, in *Proceedings of Eurographics* (2009), pp. 369–378
29. A.A. Goshtasby, Fusion of multi-exposure images. Image Vis. Comput. **23**(6), 611–618 (2010)
30. J. Herwig, J. Pauli, An information-theoretic approach to multi-exposure fusion via statistical filtering using local entropy, in *Proceedings of International Conference on Signal Processing, Pattern Recognition and Applications* (2013), pp. 50–57
31. R. Shen, I. Cheng, J. Shi, A. Basu, Generalized random walks for fusion of multiexposure images. IEEE Trans. Image Process. **20**(12), 3634–3646 (2012)
32. H. Singh, V. Kumar, S. Bhooshan, Weighted least squares based detail enhanced exposure fusion, in *Proceedings of ISNR Signal Processing* (2014), pp. 62–69
33. W.H. Cho, K.S. Hong, Extending dynamic range of two color images under different exposures, in *Proceedings of International Conference on Pattern Recognition* (2014), pp. 853–856
34. M. Song, D. Tao, C. Chen et al., Probabilistic exposure fusion. IEEE Trans. Image Process. **21**(1), 341–357 (2012)
35. C.H. Hsieh, B.C. Chen, C.M. Lin, et al., Detail aware contrast enhancement with linear image fusion, in *Proceedings of International Symposium on Aware Computing* (2010), pp. 1–5
36. L. Pei, Y. Zhao, H. Luo, Application of wavelet-based image fusion in image enhancement, In *Proceedings of International Congress on Image and Signal Processing* (2010), pp. 649–653
37. S. Raman, S. Chaudhuri, A matte-less, variational approach to automatic scene compositing, In *Proceedings of International Conference on Computer Vision (ICCV)* (2007), pp. 574–579
38. K. Kotwal, S. Chaudhuri, An optimization-based approach to fusion of multiexposure, low dynamic range images, in *Proceedings of International Conference on Information Fusion* (2011), pp. 1942–1948
39. X. Dong, G. Wang, Y.A. Pang. et al., Fast efficient algorithm for enhancement of low lighting video, in *Proceedings of IEEE International Conference on Multimedia and Expo* (2011), pp. 1–6
40. L. Li, R. Wang, W. Wang, et al., A low-light image enhancement method for both denoising and contrast enlarging, in *Proceedings of IEEE International Conference on Image Processing (ICIP)* (2015), pp. 3730–3734
41. X. Guo, LIME: A method for low-light image enhancement, in *Proceedings of ACM on Multimedia Conference* (2016), pp. 87–91
42. K.G. Lore, A. Akintayo, S. Sarkar, LLNet: a deep autoencoder approach to natural low-light image enhancement. Pattern Recognit. **61**, 650–662 (2017)
43. W. Ren, S. Liu, L. Ma et al., Low-light image enhancement via a deep hybrid network. IEEE Trans. Image Process. **28**(9), 4364–4375 (2019)
44. R. Wang, Q. Zhang, C.-W. Fu, X. Shen, W.-S. Zheng, J. Jia, Underexposed photo enhancement using deep illumination estimation, in *Proceedings of IEEE/CVF Conference on Computer Vision and Pattern Recognition (CVPR)* (2019), pp. 6842–6850
45. J. Cai, S. Gu, L. Zhang, Learning a deep single image contrast enhancer from multi-exposure images. IEEE Trans. Image Process. **27**(4), 2049–2062 (2018)
46. K. Xu, X. Yang, B. Yin, Rynson W.H. Lau, Learning to restore low-light images via decomposition-and-enhancement, in *Proceedings of IEEE/CVF Conference on Computer Vision and Pattern Recognition (CVPR)* (2020), pp. 2278–2287

47. Y. Zhang, J. Zhang, X. Guo, Kindling the darkness: a practical low-light image enhancer, in *Proceedings of the ACM International Conference on Multimedia (ACM MM)* (2019), pp. 1632–1640

48. W. Yang, S. Wang, Y. Fang, et al., From fidelity to perceptual quality: a semi-supervised approach for low-light image enhancement, in *Proceedings of IEEE/CVF Conference on Computer Vision and Pattern Recognition (CVPR)* (2020), pp. 3060–3069

49. C. Guo, C. Li, J. Guo, et al., Zero-reference deep durve estimation for low-light image enhancement, in *Proceedings of IEEE/CVF Conference on Computer Vision and Pattern Recognition (CVPR)* (2020), pp, 1777–1786

50. H. Lee, K. Sohn, D. Min, Unsupervised low-light image enhancement using bright channel prior. IEEE Signal Process. Lett. **27**, 251–255 (2020)

51. E.P. Bennett, L. Mcmillan, Video enhancement using per-pixel virtual exposures. ACM Trans. Graph. **24**(3), 845–852 (2005)

52. Q. Zhang, Y. Nie, L. Zhang et al., Underexposed video enhancement via perception-driven progressive fusion. IEEE Trans. Vis. Comput. Graph. **22**(6), 1773–1785 (2016)

53. X. Dong, L. Yuan, W. Li, et al., Temporally consistent region-based video exposure correction, in *Proceedings of IEEE International Conference on Multimedia and Expo* (2015), pp. 1–6

54. S. Liu, Y. Zhang, Detail-preserving underexposed image enhancement via optimal weighted multi-exposure fusion. IEEE Trans. Consum. Electron. **65**(3), 303–311 (2019)

55. S. Liu, Y. Zhang, Non-uniform illumination video enhancement based on zone system and fusion, in *Proceedings of International Conference on Pattern Recognition (ICPR)* (2018), pp. 2711–2716

56. C. Chen, Q. Chen, M. Do, V. Koltun, Seeing motion in the dark, in *Proceedings of IEEE/CVF International Conference on Computer Vision (ICCV)* (2019), pp. 3184–3193

57. C. Zheng, Z. Li, Y. Yang, S. Wu, Single image brightening via multi-scale exposure fusion with hybrid learning. IEEE Trans. Circuits Syst. Video Technol. **31**(4), 1425–1435 (2020)

58. D. Triantafyllidou, S. Moran, S. McDonagh, S. Parisot, G. Slabaugh, Low light video enhancement using synthetic data produced with an intermediate domain mapping, in *Proceedings of European Conference on Computer Vision (ECCV)* (2020), pp. 103–119

inted in the United States
Baker & Taylor Publisher Services